# THE
# POCKET CALCULATOR
# GAME BOOK

Now anyone can play authentic games with numbers. All you need is the greatest pocket calculating tool on earth — your own electronic marvel.

FUN FOR 1, 2, 3, 4 OR MORE

The first and only book of its kind for learning and pleasure.

# THE
# POCKET CALCULATOR
# GAME BOOK

## BY EDWIN SCHLOSSBERG AND JOHN BROCKMAN

## CORGI BOOKS
### A DIVISION OF TRANSWORLD PUBLISHERS LTD

THE POCKET CALCULATOR GAME BOOK
A CORGI BOOK 0 552 98005 6

First publication in Great Britain

Printing History
Corgi edition published 1976
Corgi edition reprinted 1976
Corgi edition reprinted 1977
Corgi edition reprinted 1978

Corgi Books are published by
Transworld Publishers Ltd.,
Century House, 61-63 Uxbridge Road,
Ealing, London W.5

Printed by James Paton Ltd., Paisley Scotland.

# THE
# POCKET CALCULATOR
# GAME BOOK

# ACKNOWLEDGMENTS

This book could not have been done without the sustained help and the contributions of Kenneth L. Wilson. His creativity and insight have been invaluable.

We owe a debt of gratitude to Professor Heinz Von Foerster. Dr. Von Foerster, an eminent cybernetician, has been, as always, instrumental in the creation of productive contexts. We thank Ricardo Uribe for the games he created. They are an asset to the book.

Special thanks to the following individuals who contributed their time and ideas: Michael Brun, Brian Davis, Andy Dorner, Medard Gabel, David Grothe, Richard Marks, Hugh McCarney, Lenore Metrick, Peter Pearce, Susan Pearce, H. A. Peele, Robert Rebitzer, Laura Selby, Jerry Sturmer, Ann Wilson, Bob Wurman, and Professor D. D. Zettel.

Vincent Trocchia is responsible for the graphics and Katinka Matson coordinated the final stages of the book. We wish to thank our editors, James Landis and Jack Looney, for their invaluable guidance.

This book was conceived by Edwin Schlossberg and developed by John Brockman. Two colleagues, Kenneth Wilson and Ricardo Uribe, created twenty-two of the fifty games and puzzles. Their contributions are listed below. The remaining twenty-eight games and puzzles were invented by Edwin Schlossberg. All the games and puzzles in the book were developed and edited by John Brockman.

Ricardo Uribe:

COLD WAR
COVERT OPERATION
DÉTENTE
MIND CONTROL
SECRET ENTERPRISE

Kenneth L. Wilson:

49er
50
1000
THE BOARD GAME
CALCULATOR POKER

CALCULATOR SOLITAIRE
CALCUMAZE
CALCUMEASURE
COMMANDER-IN-CHIEF
DICE MAZE
DOUBLE SOLITAIRE
DUO-MAZE
LOVER'S MAZE
MULTI MAX
MULTIPLICATION MAZE
QUARTET MAZE
SERIES SOLITAIRE

# CONTENTS

# THE
# POCKET CALCULATOR
# GAME BOOK

# Introduction

Pocket calculators are new to our lives. Unknown five years ago, they are becoming as popular as televisions or hi-fi sets. Yet they are different in that they are not a passive entertainment but require intelligent input and definite intention for their use. The potential uses for the pocket calculator are many and varied, but at present such use seems limited in general to business and individual economic and household situations.

We have written *The Pocket Calculator Game Book* with the following purposes in mind: (1) to provide you with interesting, exciting, and amusing experiences; (2) to create new contexts for game playing and human interaction; and (3) to develop numerous practical ways for you to use this new and valuable tool.

We have tried to go beyond merely presenting the capabilities of the pocket calculator to do mathematics. We are not so much interested in *what the pocket calculator can do* as we are in *what you can do with your pocket calculator.* Through using this book you can turn your pocket calculator into a game board, a puzzle board, a deck of cards, a maze runner, a social interaction device, and so on. The games and puzzles will involve you in situations of human competition, human interest, and human excitement.

## ABOUT THE GAMES AND PUZZLES

The games and puzzles presented in this book are as varied and different from one another as Gin Rummy is from Monopoly, as Backgammon is from Scrabble, as Parcheesi is from Checkers. We believe that many will take on lives of their own quite apart from this book, and provide you with many years of playing enjoyment.

The fifty games and puzzles in this volume are based on card games, life situations, business problems, social concerns, government, economy, familiar board games, and interesting situations. Some are based on chance, some on estimation. Some are mazes, some are explorations of methods of self-understanding.

Many of the games are cooperative in nature. They are designed to be played so that everyone wins from the experience. Many are competitive: there are winners and losers. Some provide entertaining, simple experiences. Some are complex and require concentration and deep involvement. All are accessible to anyone with the ability to operate a pocket calculator.

In many cases a particular genre of game is presented as a one-player (puzzle) version, two-player version, and three- or more-player version. In such cases, we recommend that you start with the puzzle, familiarize yourself with the play, and proceed to the multiplayer games. These groups of games, which are based on similar ideas, sometimes have very different rules of play. Be sure to read the instructions carefully for each game.

In most of the fifty games and puzzles we have provided a final paragraph outlining a Complex Version. This will make the games and puzzles more difficult, thus increasing the challenge as well as providing you with more opportunities for play once you become skilled in simpler versions.

## ABOUT USING YOUR POCKET CALCULATOR

All the games and puzzles are designed to be played with the standard four-function pocket calculator (functions are also known as operations). Instructions for each game are based on the way most calculators operate. Your calculator may be different in its way of operating, so be sure that you understand the rules of the games before you play so that you can adapt them to the way in which your particular calculator operates, if necessary. Some of the Complex Versions involve using square root and trigonometric functions as well.

We use several different words in the games and puzzles to describe operations of the calculator and ways to play. The following diagram will show you some of these:

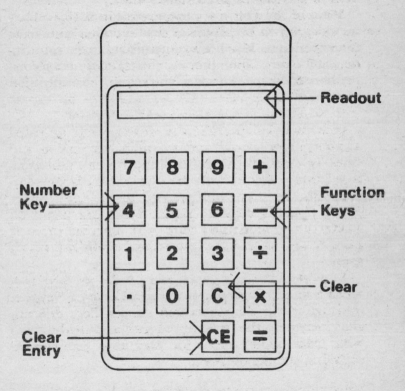

### DIGITS AND NUMBERS

Digits are like letters and numbers are like words. Words are made of letters and numbers are made from digits. For example: the number 632 is made up of the digits 6 and 3 and 2. 632 is a three-digit number.

### FUNCTIONS AND OPERATIONS

These are + (add), − (subtract), × (multiply), ÷ (divide). An operation is what you do with one number to another number. You add one number to another number, you subtract one number from another number, you multiply one number by another number, you divide one number by another number. The functions or operations refer to adding, subtracting, multiplying, and dividing.

### PLUS AND MINUS NUMBERS

A plus number is a number greater than 0 (for example: 6 is a plus number). A minus number is a number less than 0 (for example: − 6 is a minus number).

### DECIMAL POINTS AND A DECIMAL NUMBER

A decimal point is the point separating a whole number from a number less than 1. For example: if you have ten and a half apples, you will see 10.5 on the calculator. The .5 is the same as ½ and is a decimal number. A decimal number is a number less than 1. The numbers to the right of the decimal point on the calculator are the decimal numbers (for example: .5 above).

A fraction such as ½ is shown on the calculator as .5, since it is 5/10 or half of 1. Thus, when a fraction comes up in a game or puzzle, it is the same as a decimal number (a part of 1 such as 1/2 or 1/4, which is shown on the calculator as .5 or .25).

### RANDOM NUMBERS

In some of the games it is important that the player cannot choose the number that he or she plays. In such cases random numbers are generated. A random number is simply a number that we did not pick ourselves. As we did not choose it, it provides us with a nonpredictive chance element which makes the games and puzzles more interesting. In this book, we generate random numbers by dividing a five-digit number by any other five-digit number. The result of such a division is complicated enough so that it cannot be accurately predicted without the use of the pocket calculator. For example: $12345 \div 54321 = 0.2272601$. If we are generating a five-digit random number, we use the last five digits, or, 72601. If we are generating a three-digit random number, we use the last three digits, or, 601, etc.

Most games and puzzles usually challenge the players to do something that has already been figured out by the inventor. The challenge is thus not to discover something new, but to redo something that has already been worked out for you. We have attempted to create games and puzzles which are open-ended in terms of your involvement. We hope that through using this book you will open and explore new ideas, and have many long hours of productive learning and pleasure.

EDWIN SCHLOSSBERG          JOHN BROCKMAN
Chester, Massachusetts     New York City

# Even and Odd

**2 Players**
**1 or 2 Calculators**
**Paper and Pencil**

## OBJECT OF THE GAME

To reach either an even or odd number as the final number. The game can be played on one calculator, or each player can use a separate calculator.

## THE PLAY

The game is organized around the ten digits and four mathematical operations on the calculator ($+$ $-$ $\times$ $\div$). One player is the even player and one player is the odd player. They respectively try to reach an even or odd number as the final number.

The players decide who will go first. This can be done by flipping a coin. Player A chooses even or odd. If even, A tries to make the final number which appears in the calculator even; if odd, A tries to make the final number odd.

The "board" of the game is the numbers and functions on the calculator, the numbers being 0–9. Each number can be played once. Each player plays a number and function in turn. The player is trying to make the final number even or odd after the ten turns have been played. (Neither player can multiply or divide by 0.)

The first player plays a number. The second player then plays a function and a number. This has taken up two numbers. There are eight more. The players must figure out how to play the remaining numbers to reach an even or odd result. Neither player can multiply or divide by zero.

If the divide function is used, the numbers may go into fractions (example: 6.56). Such a number is considered odd since only one place in the fraction will be considered for the game (*i.e.* 6.5).

If the players find it difficult to remember which numbers have been played, they can make a list of 0–9 and cross out each number as it is played.

## SAMPLE PLAY

Players toss a coin to decide who goes first. A wins the toss and decides to be even.

| | |
|---|---|
| A plays 2 | The calculator reads 2 |
| B plays + 6 | The calculator reads 8 |
| A plays × 3 | The calculator reads 24 |
| B plays + 4 | The calculator reads 28 |
| A plays − 7 | The calculator reads 21 |
| B plays × 5 | The calculator reads 105 |
| A plays − 9 | The calculator reads 96 |
| B plays ÷ 8 | The calculator reads 12 |
| A plays × 1 | The calculator reads 12 |
| B plays + 0 | The calculator reads 12 |

### A wins

## COMPLEX VERSION

In this version, trigonometric, exponential, and logarithmic functions of more complex calculators are included.

The play thus includes more decimal point numbers and the same rule applies to rounding off. The number under consideration is the first number to the right of the decimal point. For example: 6.7843536 is odd, being considered 6.7.

# Give and Take

**2 Players**
**2 Calculators**

**OBJECT OF THE GAME**

To get your number over 999999.

**THE PLAY**

Each player enters a six-digit number in his or her calculator, no two digits of which are the same. A coin toss then decides who goes first. Player A says, "Give me your 5's." (This is an example; a player can ask for any number from 1–9.) Player B, reading his or her calculator, says, "You get 500."

The "give and take" of the above number depends on where the 5 occurs in the number of Player B. If the number on the calculator reads 12345, B says, "You get 5"; 12354, B says, "You get 50"; 12534, B says, "You get 500"; 15234, B says, "You get 5000"; et cetera.

The player who "takes" adds the value of the number. The player who "gives" subtracts the same value. Thus, Player A says, "Give me your 5's." Player B says, "Take 50." A adds 50 to his or her number; B subtracts 50.

If one player asks for a number the other player does not have (for example: A asks for 6, and B says, "I have no 6's"), the play continues with B asking A for a number.

Play continues until one player wins by going over 999999. No player can ask for 0. If a player has two or more of the same digits in his or her number, the smaller of the two numbers may be given. (For example: 663790. The player gives 60,000, not 600,000).

## STRATEGY

Putting a large or small initial number in the calculator can be very risky. As you play, numbers starting with 5 and 6 will look increasingly attractive. Also, players sometimes reveal the number they don't want to give away by asking for it.

## SAMPLE PLAY

|  |  | A | B |
|---|---|---|---|
| | Starting Number: | 765432 | 658792 |
| *Plays:* | | | |
| A: | "Give me 2" | 765432 | 658792 |
| | | + 2 | — 2 |
| | | 765434 | 658790 |
| B: | "Give me 5" | — 5000 | + 5000 |
| | | 760434 | 663790 |
| A: | "Give me 6" | + 60000 | — 60000 |
| | | 820434 | 603790 |
| B: | "Give me 8" | — 800000 | + 800000 |
| | | 20434 | 1403790 |

B wins

# Second Hand

**2 Players**
**2 Calculators**

## OBJECT OF THE GAME

To arrive at a six-digit number, the sum of whose digits is higher than that of your opponents.

## THE PLAY

Each player puts a one-digit number into his calculator. Player A then states a one-digit number. Player B enters the difference between that number and the last entered number that was on B's calculator. (If a minus number results, disregard it. For example: $6 - 9 = -3$, then player should enter 3.)

For example:

A's first number is 6.

B's first number is 9.

A plays 4.

B takes the difference $9 - 4 = 5$, and enters it as the second number. B's readout is now 95.

B plays 9.

A takes the difference $9 - 6 = 3$, and enters it as the second number. A's readout is 63.

Play continues until both players have a six-digit number on their calculators. The digits are then added (for example: $123456 = 1 + 2 + 3 + 4 + 5 + 6 = 21$). The highest number wins. Minus numbers should be disregarded.

An extended version is to play for a total score of 350 points. The first player to reach the total wins the game.

**STRATEGY**

Try to figure out your opponent's plan to develop his or her number, then do everything possible to interfere.

**SAMPLE PLAY**

|  | A |  | B |
|---|---|---|---|
| First number: | 6 |  | 9 |
| A plays 5 therefore |  | (9 — 5 = 4) Result: | 94 |
| B plays 6 therefore |  |  |  |
| (6 — 6 = 0) Result: | 60 |  |  |
| A plays 6 therefore |  | (4 — 6 = 2) Result: | 942 |
| B plays 8 therefore |  |  |  |
| (0 — 8 = 8) Result: | 608 |  |  |
| A plays 8 therefore |  | (2 — 8 = 6) Result: | 9426 |
| B plays 3 therefore |  |  |  |
| (8 — 3 = 5) Result: | 6085 |  |  |
| A plays 2 therefore |  | (6 — 2 = 4) Result: | 94264 |
| B plays 1 therefore |  |  |  |
| (5 — 1 = 4) Result: | 60854 |  |  |
| A plays 8 therefore |  | (4 — 8 = 4) Result: | 942644 |
| B plays 9 therefore |  |  |  |
| (4 — 9 = 5) Result: | 608545 |  |  |

A has $608545 = 6 + 0 + 8 + 5 + 4 + 5 = 28$
B has $942644 = 9 + 4 + 2 + 6 + 4 + 4 = 29$, so B wins.

# Calculator Solitaire

1 Player
1 Calculator
1 Deck of Cards

**OBJECT OF THE GAME**

This game is like solitaire in its aim to eliminate all the cards by pairing up or by matching suits. In this game, however, you must be able to evenly divide a card in order to eliminate it.

**THE PLAY**

Remove all face cards and jokers from the deck. Make three piles of seven cards each, and turn over the top card of each pile. Multiply the numbers on the top cards together on your calculator. Deal two cards face up. Divide one card into the number on your calculator. If it comes out even, that card can be discarded. Multiply the three top cards on the three piles again, and divide the product by the other card you turned over. If this comes out evenly, you can eliminate the top three cards and the two cards you turned over. If they do not divide evenly, leave the cards on the three piles and turn over two more cards from the deck. Continue until you have eliminated all the cards. If no pair is found that will divide evenly, you have lost the game.

An example should help to clarify:

Top Card On Each Stack   $\boxed{3}$  X  $\boxed{10}$  X  $\boxed{9}$  = 270

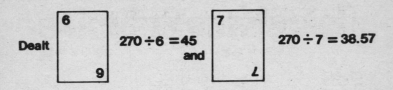

Dealt $270 \div 6 = 45$ and $270 \div 7 = 38.57$

Thus 7 does not divide into 270. Two more cards are dealt.

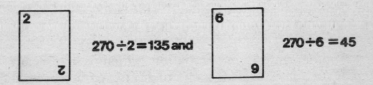

$270 \div 5 = 54$ and $270 \div 8 = 33.75$

8 does not divide into 270 and two more cards are dealt.

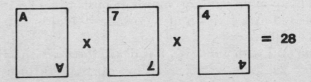

$270 \div 2 = 135$ and $270 \div 6 = 45$

Both divide evenly so all five cards (3, 10, 9, 2, 6) are set aside and the next card on each pile turned over revealing:

A X 7 X 4 $= 28$

and the two cards dealt from the pack are:

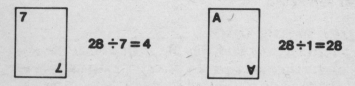

| | |
|---|---|
| **10** | $28 \div 10 = 2.8$ |
| **2** | $28 \div 2 = 14$ |

which does not work. The next two are:

| | |
|---|---|
| **7** | $28 \div 7 = 4$ |
| **A** | $28 \div 1 = 28$ |

Both divide evenly so all five are set aside. And so it continues until all three stacks are exhausted or no pair will divide the product of the stacks.

**COMPLEX VERSION**

There are many variations of this basic solitaire. For a more difficult game to win, deal stacks of eight cards each.

Another interesting variation which is a difficult game to win is to deal stacks of six (seven for an even stiffer game). When pairs are dealt from the pack, divide each card in turn into the product without returning to the original product after a division. Thus, if the stack cards are $6 \times 7 \times 4 = 168$, and the pair 8, 2 is dealt, $168 \div 8 = 21$ and $21 \div 2 = 10.5$ which is not a whole number. Note that 168 is divisible by 8 and by 2 individually, but not simultaneously. In effect, this game requires division of the product (168) by the product of the pair ($8 \times 2 = 16$).

Playing still with 168: if 7, 8 is dealt, the division is $168 \div 7 = 24$, $24 \div 8 = 3$, so all five cards are thrown out.

# Multi Max

2 Players
2 Calculators
1 Deck of Cards
Paper and Pencil

## OBJECT OF THE GAME

To rearrange the cards you are dealt so that the result of multiplying them together in pairs is larger than the number you get when multiplying them as they were dealt to you.

## THE PLAY

Take a deck of cards and eliminate the 10's, jacks, queens, kings, and jokers. Shuffle the cards and deal six cards, face up, to each player. The cards are to be arranged by the dealer into three sets of two cards each. Thus, the players have three pairs of cards. (For example: if a player is dealt 5, 2, 1, 8, 5, 9 they are paired as 52, 18, and 59.)

The player now multiplies these pairs together. This is the DEAL NUMBER. After the players calculate their DEAL NUMBERS, they are free to make a new arrangement of the cards to make a higher total when they are multiplied together. The players do this. The players multiply the new total together (for example: in the above example, the player might rearrange the number to 52, 81, 95). This is the MAX NUMBER. The players subtract their DEAL NUMBER from their MAX NUMBER and thus arrive at the score for that hand.

Additional hands are played until one player's score total reaches 1,000,000. If two players go over 1,000,000 in the same hand, the player with the highest score wins the game. Three games are needed to win a set.

## SAMPLE PLAY

Player A is dealt 2, 9; 5, ace; 4, 8 making a product $29 \times 51 \times 48 = 70,992$.

Player B is dealt 5, ace; 8, 4; 6, 7 making a product $51 \times 84 \times 67 = 287,028$.

Player A rearranges to get 9, 2; 5, ace; 8, 4 making a new product: $92 \times 51 \times 84 = 394,128$, and a resulting score $394,128 - 70,992 = 323,136$.

Player B rearranges to get 7, 5; 8, ace; 6, 4 making a new product $75 \times 81 \times 64 = 388,800$, and a resulting score $388,800 - 287,028 = 101,772$. Play continues:

| HAND/ PLAYER | CARDS AS DEALT | DEAL NUMBERS | MAX NUMBER | PRODUCT | SCORE | TOTAL |
|---|---|---|---|---|---|---|
| 2 A | 43, 13, 58 | 32,422 | 43, 81, 53 | 184,599 | 152,177 | 475,313 |
| 2 B | 47, 73, 69 | 236,739 | 94, 73, 76 | 521,512 | 284,773 | 386,545 |
| 3 A | 32, 56, 29 | 51,968 | 63, 52, 92 | 301,392 | 249,424 | 724,737 |
| 3 B | 69, 27, 18 | 33,534 | 91, 76, 82 | 567,112 | 533,578 | 920,123 |

reshuffle

| | | | | | | |
|---|---|---|---|---|---|---|
| 4 A | 18, 16, 84 | 24,192 | 81, 61, 84 | 415,044 | 390,852 | 1,115,589 |
| 4 B | 74, 84, 56 | 348,096 | 74, 84, 65 | 404,040 | 55,944 | 976,067 |

So A wins in the fourth hand, after B's strong third hand and a disappointing fourth hand.

## STRATEGY

It should be obvious that the sequence 28, 35, 69 (product $28 \times 35 \times 69 = 67,620$) can be improved greatly by merely reversing each digit of each pair (thus: 82, 53, 96, giving a product $82 \times 53 \times 96 = 417,216$). What is less obvious is the rearrangement $82 \times 93 \times 65 = 495,690$. There is an optimal strategy for rearrangement which will maximize your score. The arrangement can be found with a bit of experimentation.

**COMPLEX VERSION**

Cards are dealt from the remaining deck without reshuffling until deck is exhausted, or until a complete hand cannot be dealt. At this point all cards are reshuffled thoroughly. For two players, three hands may be dealt; three players, two hands; over three, each hand is reshuffled.

# Magic Number

**3 or More Players**
**1 Calculator per Player**
**Paper and Pencil**

## OBJECT OF THE GAME

MAGIC NUMBER is similar to a card game in that each player is trying to achieve a perfect "hand," the "hand" being a five-digit number (*e.g.* 53832) which the player selects at the outset of the game. Each play in the game is like making a bid in a card game, but in MAGIC NUMBER, the "bid" is the stating of (1) a number, and (2) a calculator operation (*i.e.* + − × ÷).

## THE PLAY

Each player selects a five-digit number and writes it down on a piece of paper. This is the MAGIC NUMBER. *The player does not show this number to the opponents.*

Next, each player selects a number from 1–10. This is the STARTING NUMBER. The players announce their STARTING NUMBERS to each other. They may pick the same or different numbers, it doesn't matter.

The player must get from the STARTING NUMBER to the MAGIC NUMBER using operations of the calculator. This is done by adding, subtracting, multiplying, or dividing on the calculator until one of the players reaches his or her MAGIC NUMBER.

Part of the excitement of the game is that each player's current number is available to the other players on the visual readout of the calculator. Thus, each player knows the status of the other players.

After all players have chosen a MAGIC NUMBER and a STARTING NUMBER, Player A takes a turn:

he or she states an operation (+ − × ÷) and a two- or three-digit PLAY NUMBER. For example: "multiply by 200"; "add 343"; "divide by 20"; et cetera. *All players must carry out Player A's command on their own STARTING NUMBERS.*

Player B then takes a turn, then Player C and so on until:

1) one of the players reaches his or her MAGIC NUMBER and wins the game

2) one of the players goes over the LIMIT NUMBER, which is 999999, and is automatically eliminated, after which play resumes until

3) all but one of the players goes over the LIMIT NUMBER, and the remaining player wins the game.

If the number of the game goes into fractions (numbers to the right side of the decimal point) the number is rounded off to the next largest whole number (*e.g.* 89.4 = 90).

**STRATEGY**

1) Players must try to guess the MAGIC NUMBER of their opponents, and keep play away from that number.

2) A player may try to make the opponents go over the LIMIT NUMBER, or may assess the MAGIC NUMBER of the other players and try to limit the numbers so the opponent is unable to reach the MAGIC NUMBER.

3) As the player becomes more experienced, it is possible for him or her to set up patterns between the MAGIC NUMBER and the STARTING NUMBER. One possibility is to make the two numbers very different in order to elude the attention of opponents.

4) Bluffing is a major factor in the game as each turn indicates where a player wants to go. Bluffing may be used to make the opponents miss the range of the MAGIC NUMBER and thereby give the bluffer free range to reach his or her MAGIC NUMBER first.

## SUMMARY OF PLAY

1) Players select MAGIC NUMBERS
    A selects 11111
    B selects 22222
    C selects 33333

2) Players write MAGIC NUMBERS down or put them into calculators' memory. They do not show them to opponents.

3) Players select starting numbers and announce them to opponents:
    A selects 1
    B selects 2
    C selects 3

4) Play begins:
    A states: + 400 (add 400)
    Thus: A = 401, B = 402, C = 403
    B states: × 22 (multiply by 22)
    Thus: A = 8822, B = 8844, C = 8866
    C states: + 277 (add 277)
    Thus: A = 9099, B = 9121, C = 9143

5) Play continues until one player reaches his or her particular MAGIC NUMBER and wins, or all but one of the players go over the LIMIT NUMBER and are eliminated.

## COMPLEX VERSION

This is a variation of MAGIC NUMBER for players who have complex calculators. The game is not more complex or difficult, but the range of play is expanded. It is recommended that players first try the basic game before attempting this version.

In this version of the game, the rules are similar to the basic game, with the following exceptions:

1) Instead of stating your number and a simple operation (+ − × ÷) as your turn, you may only state any one of the complex functions (such as exponents or logarithms).

2) The MAGIC NUMBER is seven digits instead of five digits.

3) The PLAY NUMBER can be one or two digits, instead of two or three digits.

4) The STARTING NUMBER must be selected from numbers 10 through 20.

5) The LIMIT NUMBER is 9,999,999.

The reasons for these changes is that the complex functions accelerate the rate by which a number grows, so there has to be a wider range of numbers and a smaller amount of each play so that the play becomes more intense.

# Bicentennial

**2 Players**
**2 Calculators**
**1 Pair of Dice**

## OBJECT OF THE GAME

To make your calculator read 1976 in as few moves as possible.

## THE PLAY

Each player rolls the dice and enters the sum of the two numbers in the calculator. The player with the highest starting sum begins the play.

If a player rolls 5, then 5 is entered in the player's calculator. However, if 7 or 11 is rolled, something special happens. A player rolling 7 must divide by 7; a player rolling 11 must multiply by 11. However, if the first roll is 7 or 11, simply enter 7 or 11 in your calculator. All decimals resulting from division are eliminated from the calculator immediately (for example: $130 \div 7 = 18.57142$; the number becomes 18). Any number except 7 and 11 can be added or subtracted.

The play limit to BICENTENNIAL is two hundred rolls. Very few games will go on this long. If the limit is reached, the player with the number closer to 1976 is the winner.

## SAMPLE PLAY

|        | A         |     | B          |     |
|--------|-----------|-----|------------|-----|
| Roll 1 |           | 12  |            | 5   |
| Roll 2 | $+ 4 =$   | 16  | $+ 9 =$    | 14  |
| Roll 3 | $+ 10 =$  | 26  | $\times 11 =$ | 154 |

|         | A          |      | B          |      |
|---------|------------|------|------------|------|
| Roll 4  | + 12 =     | 38   | + 10 =     | 164  |
| Roll 5  | × 11 =     | 418  | + 12 =     | 176  |
| Roll 6  | + 4 =      | 422  | + 6 =      | 182  |
| Roll 7  | × 11 =     | 4642 | × 11 =     | 2002 |
| Roll 8  | ÷ 7 =      | 663  | — 10 =     | 1992 |
| Roll 9  | + 8 =      | 671  | — 4 =      | 1988 |
| Roll 10 | ÷ 7 =      | 95   | — 12 =     | 1976 |

**B wins**

# 1001

1 Player
1 Calculator
1 Pair of Dice

**OBJECT OF THE PUZZLE**
 To get to 1001 as fast as possible.

**THE PLAY**
 Throw the dice and consult the following chart:

$$1 = +$$
$$2 = -$$
$$3 = \times$$
$$4 = \div$$
$$5 = \times$$
$$6 = +$$

One die will be used to determine the number entry. The other will be used in conjunction with the chart to determine operation entry.

 For example, if the roll was 3 and 5, you could enter 3 as the number and enter $\times$ as the function. Or, you could enter 5 as the number and enter $\times$ as the function. You have to get EXACTLY to 1001 in the fewest number of turns. Each roll of the dice means you enter one number and one operation.

**STRATEGY**
 If the dice allow it, build up slowly so you will have some choice in later rolls. Take your time in deciding which number will be played and which number will determine the function. Try to plan for the best chances for the next roll. If you get $\div$, it can serve to

your advantage as it will get you to a number from which you will easily achieve 1001.

**SAMPLE PLAY**

| Roll 1 | 6, 1 | 6 + | |
|---|---|---|---|
| Roll 2 | 5, 3 | 5 × | ( 11 ×) |
| Roll 3 | 1, 4 | 4 + | ( 44 +) |
| Roll 4 | 1, 1 | 1 + | ( 45 +) |
| Roll 5 | 4, 2 | 4 — | ( 49 —) |
| Roll 6 | 1, 4 | 4 + | ( 45 +) |
| Roll 7 | 1, 4 | 4 + | ( 49 +) |
| Roll 8 | 3, 6 | 6 × | ( 55 ×) |
| Roll 9 | 6, 5 | 6 × | ( 330 ×) |
| Roll 10 | 3, 1 | 3 + | ( 990 +) |
| Roll 11 | 6, 6 | 6 + | ( 996 +) |
| Roll 12 | 1, 5 | 5 + | (1001 )  SUCCESS! |

**COMPLEX VERSION**

If your calculator has square root and logarithm functions, play as above but use the following chart:

$$1 = +$$
$$2 = -$$
$$3 = \times$$
$$4 = \div$$
$$5 = \text{square root}$$
$$6 = \text{logarithm}$$

# High Roller

**4 Players**
**4 Calculators**
**1 Pair of Dice**

## OBJECT OF THE GAME

To get to 1002 as fast as possible.

## THE PLAY

Each player rolls the dice, and the highest roll goes first. This number is the first number Player A enters in the calculator. The second highest goes next, et cetera.

Player A's roll of the dice allows him or her to choose which function to enter. For example, if the high roll was 12, Player A would then enter 12 and choose from $+ - \times \div$. Player B then gets to choose from the three functions that were not chosen. Player C chooses from the remaining two functions, and Player D must enter the last function. Thus:

A rolls 12, decides to enter $\times$ (from $+ - \times \div$)
B rolls  8, decides to enter $+$ (from $+ - \div$)
C rolls  6, decides to enter $-$ (from $- \div$)
D rolls  5, and has to enter $\div$

At this point the entries are as follows:

$$A = 12 \times$$
$$B = \phantom{0}8 +$$
$$C = \phantom{0}6 -$$
$$D = \phantom{0}5 \div$$

The players will roll the dice again and continue as above. The first player to reach 1002 wins the game.

**STRATEGY**

Do not assume you will be high roller on every turn and get your choice of operations. Sometimes it is best to play conservatively and select addition or subtraction as your function, rather than multiplication.

**SAMPLE PLAY**

|        | A            | B           | C               | D           |
|--------|--------------|-------------|-----------------|-------------|
| Roll 1 | 8 +          | 6 —         | 5 ÷             | 9 ×         |
| Roll 2 | 4 × (= 12)   | 3 + (= 3)   | 2 — (=   2.5)   | 1 ÷ (= 9)   |
| Roll 3 | 7 + (= 84)   | 4 — (= 7)   | 8 × (= — 5.5)   | 3 ÷ (= 3)   |

Notice that the chosen function is operational on the NEXT ROLL.

**COMPLEX VERSION**

Increase the goal number from 1002 to 100,002, and roll three dice so that the entry number is a three-digit number. The game will proceed very rapidly but will probably be more unpredictable and therefore more difficult.

# The Guessing Game

**2 Players**
**2 Calculators**
**Paper and Pencil**

## OBJECT OF THE GAME

For one player to guess the number of the other player.

## THE PLAY

Player A selects a three-digit number, writes it down, and does not show it to Player B. Player B then states a one-digit number and Player A states an operation (+ − × ÷). A is trying to guide B to the number that A has written down. A can state operations BUT NOTHING ELSE. Both players enter the stated numbers and operations into their calculators to keep track of the play. Player A is called THE NUMBER MAKER and Player B is called THE NUMBER GUESSER. When the calculator shows the number that the NUMBER MAKER wrote down, he or she must say that the number has been guessed. The two players take turns at each role.

## SAMPLE PLAY

Player A is NUMBER MAKER and writes down 764 as goal number.

| A states: | B states: | = |
|---|---|---|
|  | 6 | 6 |
| × | 9 | 54 |
| × | 7 | 378 |

| *A states:* | *B states:* | = | |
|:---:|:---:|:---:|:---|
| × | 2 | 756 | |
| + | 9 | 765 | |
| − | 3 | 762 | |
| + | 2 | 764 | SUCCESS! |

## COMPLEX VERSION

Increase the written number to a five-digit number and utilize square root and logarithm as operations. Also, use two-digit numbers on each play rather than one-digit numbers. The results will be more unpredictable and thus more difficult to anticipate.

# The Diet Calculator

1 Player
1 Calculator
Paper and Pencil

**OBJECT OF THE PUZZLE**

To see the relation between your activity, diet, and weight.

**THE PLAY**

Write down the answers to the following questions:
A) How much do you weigh?
B) How tall are you?
C) How old are you?
D) Consult TABLE 1 and find your suggested weight.
E) How many hours a day do you sleep or recline?
F) How many hours a day do you spend sitting?
G) How many hours a day do you spend standing?
H) How many hours a day do you spend walking?
I) How many hours a day do you spend exercising or doing heavy work?
   *(Make sure that the total hours add up to 24.)*

If you are a woman, do the following:
J) Multiply the answer to E times 60 (calories used per hour in this activity).
K) Multiply the answer to F times 66 (calories used per hour in this activity).
L) Multiply the answer to G times 90 (calories used per hour in this activity).
M) Multiply the answer to H times 150 (calories used per hour in this activity).
N) Multiply the answer to I times 180 (calories used per hour in this activity).

O) Add the total of answers J, K, L, M, and N. This is the total number of calories that you use in an average day.

If you are a man, do the following:

P) Multiply the answer to E times 66 (calories used per hour in this activity).

Q) Multiply the answer to F times 90 (calories used per hour in this activity).

R) Multiply the answer to G times 150 (calories used per hour in this activity).

S) Multiply the answer to H times 180 (calories used per hour in this activity).

T) Multiply the answer to I times 270 (calories used per hour in this activity).

U) Add the total of answers P, Q, R, S, and T. This is the total number of calories that you use in an average day.

V) Consult TABLE 2. Find yourself on that TABLE. If you are between twenty-two and forty-five and a man, subtract 15 calories from the calorie amount of a twenty-two-year-old for every year that you are over twenty two. If you are between twenty two and forty-five and a woman, subtract 4 calories from the calorie amount of a twenty-two-year-old for every year that you are over twenty-two. If you are between forty-five and sixty-five, either man or woman, subtract 7.5 calories from the calorie amount of a forty-five-year-old for every year that you are over forty-five.

W) You now have two calorie amounts: the amount you use every day, and the amount that your body can use every day (answers to O or U, and V).

X) You also have your weight and height in comparison to what your weight should be for your height.

The puzzle is now in your hands. Make the most out

of it. There are varied and numerous DIET methods and programs available. Rarely do you have an opportunity to examine the complex relationships of weight, calories, activities, et cetera. Using your calculator and this puzzle, you should be able to get a better grasp of your own requirements and begin planning an intelligent course of action.

TABLE 1:   Suggested Weights for Heights

| | | MEDIAN WEIGHT | |
| --- | --- | --- | --- |
| *Height*<br>in. | *Men*<br>lb. | | *Women*<br>lb. |
| 60 | | | 109 ± 9 |
| 62 | | | 115 ± 9 |
| 64 | 133 ± 11 | | 122 ± 10 |
| 66 | 142 ± 12 | | 129 ± 10 |
| 68 | 151 ± 14 | | 136 ± 10 |
| 70 | 159 ± 14 | | 144 ± 11 |
| 72 | 167 ± 15 | | 152 ± 12 |
| 74 | 175 ± 15 | | |
| 76 | 182 ± 16 | | |

TABLE 2:   Adjustment of Calorie Allowances for Adult Individuals of Various Body Weights and Ages (at a mean environmental temperature of 20°C. [68°F.], assuming light physical activity)

| BODY WEIGHT<br>lb. | RMR* AT<br>AGE 22 | AGE | | |
| --- | --- | --- | --- | --- |
| | | 22 | 45 | 65 |
| *Men* | | | | |
| 110 | 1540 | 2,200 | 2,000 | 1,850 |
| 121 | 1620 | 2,350 | 2,150 | 1,950 |
| 132 | 1720 | 2,500 | 2,300 | 2,100 |
| 143 | 1820 | 2,650 | 2,400 | 2,200 |
| 154 | 1880 | 2,800 | 2,600 | 2,400 |
| 165 | 1970 | 2,950 | 2,700 | 2,500 |
| 176 | 2020 | 3,050 | 2,800 | 2,600 |
| 187 | 2110 | 3,200 | 2,950 | 2,700 |

| BODY WEIGHT | RMR* AT | AGE | | |
|---|---|---|---|---|
| lb. | AGE 22 | 22 | 45 | 65 |
| 198 | 2210 | 3,350 | 3,100 | 2,800 |
| 209 | 2290 | 3,500 | 3,200 | 2,900 |
| 220 | 2380 | 3,700 | 3,400 | 3,100 |
| *Women* | | | | |
| 88 | 1280 | 1,550 | 1,450 | 1,300 |
| 99 | 1380 | 1,700 | 1,550 | 1,450 |
| 110 | 1460 | 1,800 | 1,650 | 1,500 |
| 121 | 1560 | 1,950 | 1,800 | 1,650 |
| 128 | 1620 | 2,000 | 1,850 | 1,700 |
| 132 | 1640 | 2,050 | 1,900 | 1,700 |
| 143 | 1740 | 2,200 | 2,000 | 1,850 |
| 154 | 1830 | 2,300 | 2,100 | 1,950 |

*Resting Metabolic Rate.

# Working Numbers

**4 Players**
**4 Calculators**
**4 Dice**
**Paper and Pencil**

## OBJECT OF THE GAME

To get the highest number on the calculator after ten rounds.

## THE PLAY

The players roll one die, and the highest roll goes first. Player A then rolls four dice. The numbers on the dice are the WORKING NUMBERS. They can be used in two ways:

1. Arranging the four digits as part of two-digit numbers (for example: 3 and 2 could be considered as 32 or 23). If you choose to use the digits in this way, *you must add the two two-digit numbers.*

2. Using the digits as single numbers. If you choose this usage, *you must multiply the single digits together.*

Thus, if your working numbers are 5, 5, 3, 1, you could use them in the following way:

    a) $55 + 31 = 86$, or better $53 + 51 = 104$
    b) $5 \times 5 \times 3 \times 1 = 75$

Each player has ten turns. The result of each turn is added to the result of the previous turns. The player with the highest score after ten turns is the winner.

**SAMPLE PLAY**

|  | Player A | Player B |
|---|---|---|
| TURN 1: | 3, 4, 6, 1 | 2, 5, 5, 3 |
|  | $43 + 61 = 104$ | $52 + 53 = 105$ |
| TURN 2: | 1, 1, 2, 3 | 4, 4, 6, 3 |
|  | $21 + 31 = \phantom{0}52$ | $4 \times 4 \times 6 \times 3 = 288$ |
|  | $\underline{\phantom{00}104}$ | $\underline{\phantom{000}105}$ |
|  | $156$ | $393$ |
| TURN 3: | 1, 6, 5, 4 | 3, 1, 2, 1 |
|  | $6 \times 5 \times 4 \times 1 = 120$ | $31 + 21 = \phantom{0}52$ |
|  | $\underline{\phantom{00}156}$ | $\underline{\phantom{000}393}$ |
|  | $276$ | $445$ |

|  | Player C | Player D |
|---|---|---|
| TURN 1: | 1, 3, 5, 6 | 2, 2, 4, 5 |
|  | $1 \times 3 \times 5 \times 6 = 90$ | $42 + 52 = 94$ |
| TURN 2: | 5, 1, 2, 3 | 5, 4, 3, 2 |
|  | $51 + 32 = \phantom{0}83$ | $5 \times 4 \times 3 \times 2 = 120$ |
|  | $\underline{\phantom{00}90}$ | $\underline{\phantom{000}94}$ |
|  | $173$ | $214$ |
| TURN 3: | 5, 4, 6, 2 | 3, 2, 6, 6 |
|  | $5 \times 4 \times 6 \times 2 = 240$ | $3 \times 2 \times 6 \times 6 = 216$ |
|  | $\underline{\phantom{00}173}$ | $\underline{\phantom{000}214}$ |
|  | $413$ | $430$ |

B leads (after three turns of the ten-turn game)

**COMPLEX VERSION**

Add the following rules to the basic game rules:

1. You may only employ usage #1 if your working numbers contain four even numbers.

2. You may only employ usage #2 if your working numbers contain only odd numbers.

3. If your working number contains a 7, multiply all the single digits except 7 and divide by 7. Add the result to your number.

4. If your working number contains a 6, you must consider three of the digits as part of a three-digit number (for example: 345) and then divide that number by 6 and add the result to the previous total.

# Divide and Conquer

**3 or more Players**
**1 Calculator per Player**

## OBJECT OF THE GAME

To have the lowest number on your calculator after six turns.

## THE PLAY

Each player enters a one-digit number and does not show it. Proceeding clockwise, each player states a one-digit number (not 0). If a player has that digit as the final digit (farthest to the right) in his or her number, the player divides by the stated number. Otherwise, the player must multiply his or her number by the stated number. However, if player has the stated number but NOT as the final digit, he or she may add it to or subtract it from the number.

Thus, if Player A states 6 and Player B has 236, Player B divides by 6 ($236 \div 6 = 39.3333$). If A states 6 and B has 235, B multiplies by 6 ($235 \times 6 = 1410$). If A states 4 and B has 1410, then B can add 4 or subtract 4 (either 1414 or 1406). Discard all decimals when reentering the number (39.3333 becomes 39).

Play continues for six turns. Whoever has the lowest number after six turns wins that round. Winning six rounds wins the game.

If there are more than three players, the number of turns is always equal to twice the number of players. However, if there are six players, each player takes one turn for a total of six turns. If there are more than six players, each player still takes only one turn.

## STRATEGY

You will run into trouble if 0 is your final digit. Try to avoid it.

## SAMPLE PLAY

|  |  | A | B | C |
|---|---|---|---|---|
| STARTING NUMBER: | | 8 | 5 | 3 |
| | STATED NUMBER | | | |
| *Player A:* | 8 | $8 \div 8 = 1$ | $5 \times 8 = 40$ | $3 \times 8 = 24$ |
| *Player B:* | 4 | $1 \times 4 = 4$ | $40 + 4 = 44$ | $24 \div 4 = 6$ |
| *Player C:* | 6 | $4 \times 6 = 24$ | $44 - 6 = 38$ | $6 \div 6 = 1$ |
| *Player A:* | 4 | $24 \div 4 = 6$ | $38 + 4 = 42$ | $1 \times 4 = 4$ |
| *Player B:* | 2 | $6 \times 2 = 12$ | $42 \div 2 = 21$ | $4 \times 2 = 8$ |
| *Player C:* | 8 | $12 \times 8 = 96$ | $21 \times 8 = 168$ | $8 \div 8 = 1$ |

C wins round

## COMPLEX VERSION

If all the players have calculators with eight or more digits, then do not discard decimal numbers. This will make the game more challenging and more unpredictable. The final digit will now be the final digit of the decimal (if, indeed, there is a decimal).

# 1000

**2 Players**
**2 Calculators**
**Paper and Pencil**

## OBJECT OF THE GAME

To reach 1000 at the end of a round.

## THE PLAY

The players enter any number in their calculators and write a second number on a sheet of paper. The players cannot see each other's numbers. Player A then instructs Player B to perform an operation between the number on B's calculator and the number on B's paper. Player A can instruct Player B to either add, subtract, multiply, or divide. B simultaneously gives similar instructions to A.

Thus, A enters 500 on the calculator and writes 2 on paper. B enters 900 on the calculator and writes 100 on paper. B instructs A to add. A instructs B to subtract: A: $500 + 2 = 502$; B: $900 - 100 = 800$.

The player who hits 1000 wins the game. If both players hit 1000, the game is a draw. If neither player hits 1000, another round is played.

In succeeding rounds, players use the results of the previous round remaining in the calculator. The players write down an additional number and instruct each other as to the operation to be performed. The game proceeds with successive rounds until a win or a draw. Players keep score by noting the number of games they win.

## SAMPLE PLAY

1. Players each enter a number in their calculators; A enters 500, B enters 100.

2. They write a second number on paper and instruct each other as to an operation. Thus:

| A WRITTEN # | A OPERATION | = | A TOTAL |
|---|---|---|---|
| 2 | + | = | 502 |
| 2 | × | = | 1,004 |
| 4 | + | = | 1,008 |
| 8 | + | = | 1,016 |
| 16 | × | = | 16,256 |

| B WRITTEN # | B OPERATION | = | B TOTAL |
|---|---|---|---|
| 10 | — | = | 90 |
| 10 | × | = | 900 |
| 10 | × | = | 9,000 |
| 8,000 | + | = | 17,000 |
| 16,000 | — | = | 1,000 |

B wins

## COMPLEX VERSION

Only two-digit numbers may be used, for both the original calculator entry and numbers written down on paper. All other rules remain the same.

# Either/Or

**1 Player**
**1 Calculator**
**1 Deck of Cards**

**OBJECT OF THE PUZZLE**

To have the largest number after ten turns.

**THE PLAY**

Remove all face cards and jokers from the deck. Shuffle and lay out the first ten cards face down corresponding to the physical pattern of number keys on your calculator.

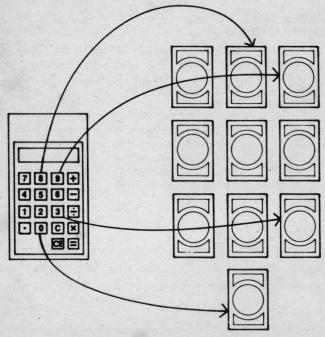

Each card corresponds to a number on your calculator. In ten turns you try to reach the largest number you can. You have the following choice on each turn: (1) use the number on the calculator; or (2) use the number of the face-down card corresponding to that number. You must choose; you cannot have both.

Each chosen number is *added* to the number already in the calculator. The play consists of playing every number or corresponding card.

A score below 45 does not count as 45 is the total of the ten calculator numbers and thus a score that can be reached with no risk by taking the cards.

If you would like to rate yourself:

$$50 = ok$$
$$70 = good$$
$$80 = very good$$
$$90 = excellent$$

## SAMPLE PLAY

|  | CARD | CALCULATOR |
|---|---|---|
| Turn 1: | — | 9 = 9 |
| Turn 2: | — | 8 = 17 |
| Turn 3: | — | 7 = 24 |
| Turn 4: | — | 6 = 30 |
| Turn 5: | 3 (did not enter 5) | — = 33 |
| Turn 6: | 8 (did not enter 4) | — = 41 |
| Turn 7: | 10 (did not enter 3) | — = 51 |
| Turn 8: | 1 (did not enter 2) | — = 52 |
| Turn 9: | 5 (did not enter 1) | — = 57 |
| Turn 10: | 7 (did not enter 0) | — = 64 |

TOTAL = 64  (ok)

## COMPLEX VERSION

Discard all four 10's from the deck of cards. The range is the same between the numbers on the calculator and the numbers on the cards.

# Calculator High

**2 Players**
**2 Calculators**
**1 Deck of Cards**

## OBJECT OF THE GAME

To reach a higher number than your opponent.

## THE PLAY

Remove all face cards and jokers from the deck. Shuffle and lay out the first ten cards face down, corresponding to the physical pattern of number keys on your calculator.

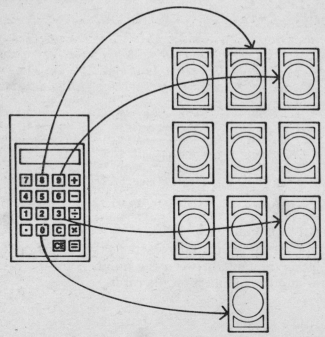

Each card corresponds to a number on the calculator. When you play the game you can choose either the card or the corresponding number on the calculator.

The players take turns choosing either to enter the number from the calculator or to risk turning over the corresponding card for that position and entering that value in the calculator. If one player chooses the number, then the other player gets the card value. If one player chooses to turn over a card, the other enters the number corresponding to that position on the calculator.

Each turn is added to the preceding turn so that after ten turns, the total on each of the calculators is the sum of the plays by each player. The winner must score over 45 for the game to count. The players draw cards to see who goes first. High card begins.

## SAMPLE PLAY

Players take turns and have the choice on their turn of either entering a number or turning over the corresponding card. Once one player chooses (*i.e.* by turning over a card), the other player must execute the remaining option (*i.e.* by entering the corresponding number) to complete the turn.

|  | | A | Total | | B | Total |
|---|---|---|---|---|---|---|
| Turn 1: | A enters | 9 | 9 | B card | 5 | 5 |
| Turn 2: | B card | 4 | 13 | A enters | 4 | 9 |
| Turn 3: | A card | 5 | 18 | B enters | 8 | 17 |
| Turn 4: | B enters | 7 | 25 | A card | 10 | 27 |
| Turn 5: | A enters | 6 | 31 | B card | 5 | 32 |
| Turn 6: | B enters | 2 | 33 | A card | 8 | 40 |
| Turn 7: | A enters | 5 | 38 | B card | 9 | 49 |
| Turn 8: | B enters | 0 | 38 | A card | 6 | 55 |
| Turn 9: | A card | 8 | 46 | B enters | 3 | 58 |
| Turn 10: | B enters | 1 | 47 | A card | 7 | 65 |

B wins

**COMPLEX VERSION**

Discard all four 10's from the deck of cards. The range is thus the same between the numbers on the calculator and the numbers on the cards.

# Calculator 21

**4 Players**
**4 Calculators**
**1 Deck of Cards**
**Paper and Pencil**

## OBJECT OF THE GAME

For each team of two players to win the hand by getting twenty-one points before the other team.

## THE PLAY

Remove all face cards and jokers from the deck. Shuffle and lay out two sets of the first ten cards, face down, corresponding to the physical pattern of number keys on the two calculators.

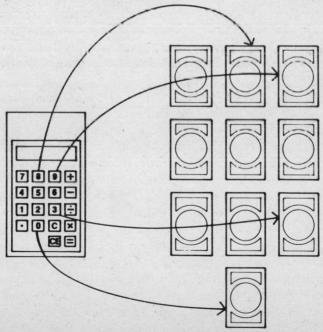

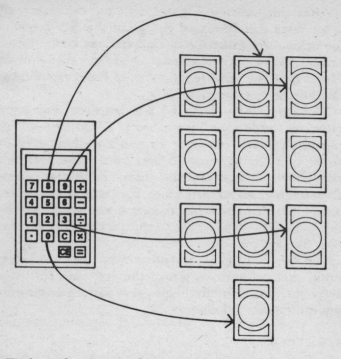

Each card corresponds to a number on the calculator. Lay out four more cards face down. The game is ready to be played.

The four players separate into two teams. The teams each play with one set of the face-down cards corresponding with the numbers on the calculator. All four players take a turn choosing whether to take the number value of the face-down card, or the number value on the calculator at that position. Before this begins, however, each team must make a guess about the total that they will have on their calculators after they have played all ten cards and numbers. Once the teams make bids, play begins. The first player takes either the number value from the calculator (for example: choosing to enter 9 rather than risking turning over the card) or decides to turn over the card corresponding to the position of 9 on the calculator, and enter that value.

After this number is entered in the calculator, the other team member must enter the number based on the option not exercised in the previous move (for example: Player A elects to enter 9 in the calculator and Player B of the same team turns up the corresponding card whose number value is 6).

The teams play until all ten numbers and corresponding cards have been played. When this is completed, the teams compare the totals of their numbers to the bids that were made. The team whose total (the cumulative number in the team's calculator) comes closest to the bid gets one point. Each player now selects one of the four cards placed face down and multiplies this times the number on their calculator. The team with the highest total number after multiplying gets one point. The game continues with new hands (two sets of ten cards each corresponding to the calculator keys and four cards for the final play) until one team gets twenty-one points to win.

### STRATEGY

A team can try to maximize their calculator number in order to win the hand when the four cards are turned over. A team can try to equalize the numbers on both calculators so that each team has an equal chance of winning the final high number. A team can try to get as close as possible to their bid and risk losing the final high number draw.

A team can try to have approximately the same number so that they will equalize their chance of having a winning number when multiplying. Or, they can risk having one player of the team have a higher number than the other which might maximize their results when multiplying at the end.

### SAMPLE PLAY

Players take turns and have the choice on their turn of either entering a number or turning over the corresponding card. Once one player chooses (*i.e.* by turning over a card), the other player must execute the remaining option (*i.e.* by entering the corresponding number) to complete the turn.

### Team 1 : Bid 98

*indicates which player's turn it is

|  |  | A |  | B |
|---|---|---|---|---|
| TURN 1: | *enters | 9 | card | 5 |
| TURN 2: | card | 4 | *enters | 4 |
| TURN 3: | *card | 3 | enters | 8 |
| TURN 4: | enters | 7 | *card | 10 |
| TURN 5: | *enters | 6 | card | 3 |
| TURN 6: | enters | 2 | *card | 8 |
| TURN 7: | *enters | 5 | card | 9 |
| TURN 8: | card | 6 | *enters | 0 |
| TURN 9: | *card | 8 | enters | 3 |
| TURN 10: | card | 2 | *enters | 1 |
|  |  | 52 |  | 51 |

A draws $6 \times 52 = 312$
B                      draws $5 \times 51 = 255$
$A + B = 312 + 255 = 567$
Bid 98 / Got $51 + 52 = 103$

### Team 2 : Bid 102

|  |  | C |  | D |
|---|---|---|---|---|
| TURN 1: | *card | 5 | enters | 4 |
| TURN 2: | enters | 7 | *card | 8 |
| TURN 3: | *enters | 9 | card | 6 |
| TURN 4: | card | 10 | *enters | 0 |
| TURN 5: | *enters | 1 | card | 6 |
| TURN 6: | enters | 2 | *card | 9 |
| TURN 7: | *enters | 5 | card | 7 |
| TURN 8: | enters | 8 | *card | 5 |
| TURN 9: | *card | 4 | enters | 6 |
| TURN 10: | card | 7 | *enters | 3 |
|  |  | 58 |  | 54 |

C draws $7 \times 58 = 406$
D                      draws $5 \times 54 = 270$
$C + D = 406 + 270 = 676$
Bid 102 / Got $58 + 54 = 112$

Team 1 has the closest bid (5 away). Team 2 has the highest number (676). Thus, Team 1 gets one point for bid and Team 2 gets one point for highest number.

The game continues until one team reaches 21 points.

# Digit

**1 Player**
**1 Calculator**

## OBJECT OF THE PUZZLE

To arrive at the largest possible number by adding up the digits of that number.

Every number has digits which have a position in that number (for example: in the number 567, 7 has the first position, 6 has the second position, 5 has the third position). Every digit also has a value independent of its value in the number (for example: In the number 567, 7 has a value of 7, 6 has a value of 6 (not 60), 5 has a value of 5 (not 500). Adding up the value of the number, you get 7 + 6 + 5 = 18).

## THE PLAY

Enter a three-digit number on the calculator with no two digits the same and no 0's. Multiply it by another three-digit number with no two digits the same and no 0's. This is the WORKING NUMBER. Notice the digits in each position (for example, if you have 22345, 5 is in the first position, 4 is in the second position, et cetera). To enlarge your number, you can do two operations:

(1) multiply the digit by its position number, or
(2) add the digit value to itself

Example of (1): If you have 123456, you can multiply 6 × 1 = 6, 5 × 2 = 10, 4 × 3 = 12, 3 × 4 = 12, 2 × 5 = 10, 1 × 6 = 6; and your number will be 713306.

Example of (2): If you have 123456, you can add 6 + 6 = 12, 5 + 5 = 10, 4 + 4 = 8, 3 + 3 = 6, 2 + 2 = 4, 1 + 1 = 2, and your number will be 246912. Notice that when you obtain a number over 9, the 1 is transferred to the digit to the left. (12 becomes 2 and 10 becomes 11; then 11 becomes 1 and 8 becomes 9; thus the last three digits in the second example are 9, 1, and 2. This is a shortcut for the foolproof method shown in the Sample Play).

Example (1): $7 + 1 + 3 + 3 + 0 + 6 = 20$
Example (2): $2 + 4 + 6 + 9 + 1 + 2 = 24$

If the number extends beyond six digits, ignore the millions digit: 1948338 becomes 948338.

You must decide which of the two operations to perform ON EACH DIGIT IN ORDER TO GET THE HIGHEST NUMBER WHEN THE DIGITS ARE ADDED.

### SAMPLE PLAY

Enter 345 and multiply it by 654 = 225630—the WORKING NUMBER.

```
1st position play 0      (no play)              =  0
2nd position play 3 × 2 (position number)     =  6
3rd position play 6 × 3 (position number)     = 18
4th position play 5 × 4 (position number)     = 20
5th position play 2 | 2 (digit value)         =  4
6th position play 2 + 2 (digit value)         =  4
```

The number becomes 461860, arrived at as follows:

```
400000 (6th position)
 40000 (5th position)
 20000 (4th position)
  1800 (3rd position)
    60 (2nd position)
     0 (1st  position)
```
—————————————
$461860 / 4 + 6 + 1 + 8 + 6 + 0 = 25$

(Note that playing $2 \times 5 = 10$ in the fifth position would give you a higher six-digit number, 521860, but its digits would only add up to 22.)

**COMPLEX VERSION**

Before selecting your WORKING NUMBER, decide on which method of calculation you plan to use: multiplying by position number, or adding digit value to digit value. Select WORKING NUMBER and carry out the chosen operation. Then carry out the other operation and see which is higher. Score yourself on the difference. Any score over 50 points is excellent.

# Position

**2 Players**
**2 Calculators**

## OBJECT OF THE GAME

To arrive at the largest possible number by adding up the digits of that number.

Numbers have two qualities: they can either indicate an amount like two apples, or they can indicate a position in a series like second place. Both qualities are used in this game.

## THE PLAY

The players enter a six-digit number in the calculator composed only of the digits 1, 2, 3, 4, 5, 6, and no more than two of the same digit (*e.g.* 665544 is all right, but 666666 is NOT). Each player's turn consists of picking a position, announcing the digit in that position, and taking the opponent's digit whose position has the same number as the digit you announced; for example: if you have 542356 during the first round of play, and you are playing on the digit in the first position (6), you can either decide to keep 6, or you can ask your opponent for the digit in the OPPONENT'S SIXTH POSITION. If your opponent has 345621, you will get 3. Your opponent will then subtract 300000 from his or her calculator and you will add 3 to your calculator. If you have a 0 in any position, you cannot ask for any number from your opponent (for example: if you have 546023, you cannot change your number in the third position). Thus, your strategy should be to avoid getting a 0 in any position at any time.

The players can play the same position number as

many times as desired. However, one player can stop the game if he or she believes the number reached is as high as possible. The players then compare numbers by adding up the digits. The player with the highest digit sum gets the difference between that sum and the opponent's sum as his or her score. Thus,

Player A has 456453    $4 + 5 + 6 + 4 + 5 + 3 = 27$

Player B has 656543    $6 + 5 + 6 + 5 + 4 + 3 = 29$

Player B gets 2 points.

The first player to reach 50 points wins the game. If a player has a number over 6 in any position THAT NUMBER CANNOT BE PLAYED BUT IT CAN BE EXCHANGED IF CALLED FOR (for example: in 555689, 8 and 9 are fixed).

## STRATEGY

In some cases it is better to stay with a number such as 6 or 7, as adding another number to it may result in reaching a 0, thus giving you no value for that position. Also, try to disregard the fact that you give and take numbers of small and large value (*e.g.* exchanging 400,000 and 40). THEIR REAL VALUE IN THIS GAME IS ONLY THEIR DIGIT VALUE.

## SAMPLE PLAY

Player A's number = 665544

Player B's number = 546456

| | A | B |
|---|---|---|
| PLAY 1: | Working on position 1, A states, "Give me your digit value of position 4": | B states, "6": |
| | 665544 | 546456 |
| | + 6 | — 6000 |
| | 665550 | 540456 |

| A | B |
| --- | --- |

PLAY 2:

| A states, "5": | Working on position 3, B states, "Give me your digit value of position 4": |
| --- | --- |

| 665550 | 540456 |
| — 5000 | + 500 |
| 660550 | 540956 |

PLAY 3:

| Working on position 3, A states, "Give me your digit value of position 5": | B states, "4": |
| --- | --- |

| 660550 | 540956 |
| + 400 | — 40000 |
| 660950 | 500956 |

PLAY 4:

| A states, "6": | Working on position 2, B states, "Give me your digit value of position 5": |
| --- | --- |

| 660950 | 500956 |
| — 60000 | + 60 |
| 600950 | 501016 |

PLAY 5: A states, "Stop the game."

| Total: | Total: |
| --- | --- |
| $6+0+0+9+5+0-20$ | $5+0 \mid 1 \mid 0 \mid 1 \mid 6-13$ |

A gets seven points $(20 - 13 = 7)$

## COMPLEX VERSION

Each player can only work on each position once. Thus, each player has only one chance to optimize the number in any position.

# Calculator Poker

**3 or more Players**
**1 Calculator per Player**
**Paper and Pencil**

## OBJECT OF THE GAME

The game is similar to Five-Card and Draw Poker and has the same object: to get the best hand and win. To familiarize yourself with the game, play Five-Card Calculator Poker first and then play Five-Card Draw Calculator Poker.

## THE PLAY

Each player enters a five-digit number and passes the calculator to the right. The next player enters the divide (÷) operation and another five-digit number (no 0's) and passes the calculator back to the player who had it originally. The CALCULATOR POKER HAND is the first five digits of this number to the right of the decimal point (for example: 53621 ÷ 56321 = .9520605; thus, 95206 is the CALCULATOR POKER HAND). Do not divide by the same number as the first entry. This will result in a hand of 1 and the process will have to be repeated.

The players consult the chart below. The Hands are listed in order. Hand A is higher than Hand B is higher than Hand C, et cetera. Aces = 1.

| | Hands | Example |
|---|---|---|
| A) | 5 numbers the same | 55555 |
| B) | Odd straight flush | 35179 |
| C) | Even straight flush | 04826 |
| D) | 4 numbers the same | 44144 |

| Hands | Example |
|---|---|
| E) Full house | 22333 |
| F) Flush: all odd or even numbers | 24448 |
| G) Straight: any 5 numbers in sequence | 12345 |
| H) Three of a number | 33318 |
| I) Two pairs of numbers | 23723 |
| J) One pair of numbers | 11356 |
| K) High number | |

If there is a tie where both players have Hands A or D or E or F or H or I or J, the highest numbers in the calculator wins (for example, 66666 wins over 55555).

If you want to bet on your calculator poker game, play Draw Calculator Poker. Play for Draw Calculator Poker differs in that once a player sees the hand and looks at the chart, he or she may bet and may ask for another 1, 2, or 3 numbers after the bidding is over.

The player turns to the player on the right and states: "I'll take the numbers you have in your third and fourth position." Player A then subtracts the numbers he or she has in these positions. Player B (the player on the right) states the numbers in these positions. Player A then enters the numbers in the calculator. Player B DOES NOT SUBTRACT THEM.

Another round of betting takes place after the discard numbers.

After this round of betting, players may discard again. However, on this subsequent round ONLY ONE number may be discarded. In this round, each player TURNS TO THE PLAYER ON HIS LEFT and asks for numbers in certain positions. For example: Player A has 23798 and says to the player on the right, "Give me your numbers in positions 4 and 5." The player on the right has 78568, so he or she would say, "5 is 7 and 4 is 8." A then subtracts 2 and 3 from his or her hand and adds 7 and 8. The hand now reads 78798.

In the second round of discarding, Player A turns to the player on the left and says, "Give me your number in the second position." The player on the left has 56789 and says, "Second is 8." A subtracts the 9 and enters the 8, and now has the following number: 78788 (a full house).

## STRATEGY

You can risk discarding the high card in order to get better numbers, and to find out what the other players have in their calculators. You should bet according to your number and what you think you can get in discarding.

## SAMPLE PLAY

Five-Card Calculator Poker:

> A  enters 34567
> B  divides it by 56321
> =  .*6137*497   61374 is hand
> B  enters 78965
> C  divides it by 56328
> =  1.*4018*78   40187 is hand
> C  enters 14896
> A  divides it by 85693
> =  .*1738*298   17382 is hand

B has winning hand: High card 874

Five-Card Draw Poker:

| A | B | C |
|---|---|---|
| enters 34567 | enters 54893 | enters 56789 |
| B divides it by 56987 = 60657 | C divides it by 56897 = 96477 | A divides it by 56123 = 01186 |
| bets | bets | bets |
| discards 3 (0, 5, 7) | discards 3 (9, 6, 4) | discards 3 (0, 8, 6) |
| asks for numbers in 1st, 2nd, 4th place from Player B, | asks for numbers in 3rd, 4th, 5th place from Player C, | asks for numbers in 1st, 2nd, 5th place from Player A, |

| A | B | C |
|---|---|---|
| gets 6, 7, 7, now has 66677* | gets 0, 1, 1, now has 01177 | gets 6, 5, 7 now has 61177 |
| bets | bets | bets |
| stays (does not discard) | discards 1 (0) asks for number in 5th place from Player A, gets 6, now has 61177 | discards 1 (6) asks for number in 5th place from Player B, gets 6, now has 61177 |
| A has full house | B has 2 pair | C has 2 pair |

A has winning hand: Full House

*Notice that C draws on A's number AFTER Player A has discarded and asked for B's numbers.

# 654321

**1 Player**
**1 Calculator**
**1 Pair of Dice**
**Paper and Pencil**

## OBJECT OF THE PUZZLE
To get the calculator to read 654321.

## THE PLAY
Enter in your calculator a two-digit number, an operation ($+ - \times \div$), and another two-digit number (for example: $23 + 34$). Next, roll a pair of dice. You can use the result of the roll any way you want, either each digit as a single number, or both digits as a two-digit number. The result is either multiplied by or divided into the above number.

Thus, if you roll 2 and 3, you can count the result as $2 + 3 = 5$, or $2 \times 3 = 6$, or 23, or 32; then you multiply $23 + 34 = 57$ by 5, 6, 23, or 32, or you divide 57 by 5, 6, 23, or 32. You must do the first part of the puzzle (enter the two-digit number, an operation, and the second two-digit number) BEFORE you roll the dice.

Each turn is either added to or subtracted from the preceding turn. Thus, if you got 45600 on the first turn and 34560 on the second turn, you can either add them ($45600 + 34560$) or subtract them ($45600 - 34560$).

You are trying to get to 654321 in the FEWEST possible turns. You must arrive at 654321 EXACTLY.

Rating: 10 turns = very good
5 turns = excellent

## STRATEGY
It will be difficult to get to 654321 if your number grows too quickly; this will force you to divide and

subtract, which is risky. Try to increase your number gradually.

## SAMPLE PLAY

| | 1ST # | OPERATION | 2ND # | DICE | OPERATION | | NEW TOTAL |
|---|---|---|---|---|---|---|---|
| Turn 1: | 25 | × | 25 | 6, 4 | × (6 + 4 = 10) | | = 6250 |
| Turn 2: | 50 | × | 50 | 5, 5 | × 10 = 25000 | + 6250 | = 31250 |
| Turn 3: | 30 | × | 45 | 3, 3 | × 33 = 44550 | + 31250 | = 75800 |
| Turn 4: | 90 | × | 90 | 5, 6 | × 56 = 453600 | + 75800 | = 529400 |
| Turn 5: | 55 | × | 25 | 4, 6 | × 64 = 88000 | + 529400 | = 617400 |
| Turn 6: | 50 | × | 50 | 3, 2 | × 23 = 57500 | + 617400 | = 674900 |
| Turn 7: | 45 | × | 45 | 5, 2 | × 10 = 20250 (subtract) | | = 654650 |
| Turn 8: | 10 | × | 10 | 3, 1 | × 3 = 300 (subtract) | | = 654350 |
| Turn 9: | 10 | × | 10 | 2, 2 | ÷ 4 = 25 (subtract) | | = 654325 |
| Turn 10: | 10 | × | 10 | 2, 5 | ÷ 25 = 4 (subtract) | | = 654321 |

**COMPLEX VERSION**

Give yourself a time limit and a maximum number of turns (for example: five minutes and fifteen turns). If your calculator has square roots and logarithms, you can add these operations into the puzzle for unpredictable results.

# Final Number

**2 Players**
**2 Calculators**
**1 Pair of Dice**
**Paper and Pencil**

**OBJECT OF THE GAME**

For the players to get the same final number.

**THE PLAY**

No talking is allowed in this game. Each player starts the game by entering a two-digit number in the calculator with both digits different and no digit being 0. Each player writes down a second two-digit number, both digits different and no digit being 0. (This number is visible to both players.)

Each player rolls one die. They must each combine the three numbers (one in the calculator, one on paper, and the number on the die) so that their FINAL NUMBERS in the calculator readouts are the same (or as close as possible).

| | CALCU-LATOR # | WRIT-TEN # | DIE # | OPERATION | FINAL NUMBER |
|---|---|---|---|---|---|
| *Player A:* | 34 | 25 | 6 | $34 + 25 + 6 =$ | 65 |
| *Player B:* | 65 | 15 | 2 | $65 + 15 \times 2 =$ | 160 |

The players are far apart. They must try to see how they are each using their numbers and try to get closer together. This is a cooperative game. The players win when each begins to understand how the other plays. The players can use any operations $(+ - \times \div)$ to arrive at their FINAL NUMBER.

## STRATEGY

One player should try to decrease the number while one increases it, but neither should try to do it drastically.

## SAMPLE PLAY

| | CALCU-LATOR # | WRIT-TEN # | DIE # | OPERATION | FINAL NUMBER |
|---|---|---|---|---|---|
| *Player A:* | 24 | 25 | 5 | $24 + 25 \times 5 =$ | 245 |
| *Player B:* | 56 | 13 | 2 | $56 \times 13 \times 2 =$ | 1456 |

(they are far apart)

| | | | | | |
|---|---|---|---|---|---|
| *Player A:* | 89 | 47 | 6 | $89 \times 47 \times 6 =$ | 25098 |
| *Player B:* | 21 | 25 | 3 | $21 + 25 \times 3 =$ | 138 |

(they are farther apart)

| | | | | | |
|---|---|---|---|---|---|
| *Player A:* | 45 | 17 | 5 | $45 + 17 \times 5 =$ | 310 |
| *Player B:* | 98 | 98 | 3 | $98 + 98 \times 3 =$ | 588 |

(they are getting closer)

| | | | | | |
|---|---|---|---|---|---|
| *Player A:* | 43 | 17 | 4 | $43 + 17 \times 4 =$ | 240 |
| *Player B:* | 61 | 64 | 2 | $61 + 64 \times 2 =$ | 250 |

(they are quite close)

Game continues until players have the same number as the FINAL NUMBER.

## COMPLEX VERSION

Use a pair of dice instead of one die. This increases the range of possibilities and makes anticipating the play increasingly difficult.

# Maze Runner

**4 Players**
**4 Calculators**
**1 Pair of Dice**
**Paper and Pencil**

## OBJECT OF THE GAME

For the MAZE RUNNER team to make it through the MAZE created by the MAZE MAKER team.

## THE PLAY

This game is for two teams of two players each. The game starts with each player entering a two-digit number in her or his calculator with the two digits different and no 0. Each player writes a two-digit number on a piece of paper which everyone can look at. The four players roll dice and the two highest rolls are Team 1. The two lowest rolls are Team 2. Team 1 is the MAZE RUNNERS. Team 2 is the MAZE MAKERS.

The players execute the first step in the maze running. Each player rolls two dice. Team 1 then tries to make a total number (THE MAZE NUMBER) less than the total number of the MAZE MAKERS (THE MAZE). To make this number, each player takes the two-digit calculator number, the two-digit written number, and the dice number (also a two-digit number), and using two operations ($+ - \times \div$) makes a total number. Then the two players of Team 1 add their total numbers together to make the MAZE NUMBER, and the two players of Team 2 add their total numbers together to make the MAZE.

THE PLAYERS MUST TAKE THE WRITTEN NUMBER AND THE CALCULATOR NUMBER

AND A FUNCTION, AND ENTER THIS VALUE
BEFORE ROLLING THE DICE. NO TALKING
ALLOWED UNTIL AFTER CALCULATIONS.

The two teams compare numbers. If the MAZE
NUMBER is smaller than the MAZE, then the MAZE
has been run and Team 1 gets 1 point. If the MAZE is
smaller than the MAZE NUMBER, then the MAZE
RUNNER has been fooled and Team 2 gets 1 point.

However, if the MAZE plus the MAZE NUMBER
do not equal 500, the MAZE IS NOT LEGITIMATE,
and neither team gets any points. Therefore, even
though each team is trying for the lowest number, they
cannot make it too low.

The first team to get 21 points wins the game.

**STRATEGY**

It is probably a good idea for a team to begin high
in their number so they can get a sense of the other
team's strategy and method of play. Also, in this way
a team can be sure that the maze will be legitimate and
that they can get some points. In considering what
number to write down and what number to enter in
the calculator, each player should be cautious. All
players can see the written number, but only you can
see your calculator number. Bluffing obviously can be
used by putting in a small written number and a very
large calculator number.

**SAMPLE PLAY**

| | CALCU-LATOR # | WRIT-TEN # | DICE # | OPERATION | |
|---|---|---|---|---|---|
| *Team 1 Maze Runners* | | | | | |
| *Player A:* | 24 | 25 | 5 | $24 + 25 \times 5 =$ | 245 |
| *Player B:* | 56 | 13 | 2 | $56 + 13 \times 2 =$ | 138 |
| | | | | MAZE NUMBER | 383 |

| | CALCU-<br>LATOR # | WRIT-<br>TEN # | DICE # | OPERATION |
|---|---|---|---|---|
| *Team 2 Maze Makers* | | | | |
| *Player C:* | 89 | 47 | 6 | $89 + 47 \times 6 = 816$ |
| *Player D:* | 23 | 34 | 11 | $23 + 34 + 11 = 68$ |
| | | | | MAZE 884 |

MAZE MAKER NUMBER + MAZE RUNNER NUMBER —
1267 (MAZE is therefore LEGITIMATE)

MAZE MAKER NUMBER — MAZE RUNNER NUMBER =
501 (MAZE RUNNER TEAM 1 gets 1 point)

## COMPLEX VERSION

Raise the number by which the maze is declared not legitimate to 750, but make a top range of 2000 over which the total MAZE NUMBER and the MAZE cannot exceed. Thus, if the total goes over 2000 or under 750 the maze is not legitimate and no points are granted.

# Traveling Salesperson

1 Player
1 Calculator
Traveling Salesperson's Maps
Paper and Pencil

**OBJECT OF THE PUZZLE**

To find the fastest and shortest routes for your sales trip.

**THE PLAY**

Look at the Traveling Salesperson's Maps. On one map are time numbers (the amount of time needed to drive from the city at one end of the line to the city at the other end of the line). On the other map are distance numbers (the distance between the two cities at the ends of each line).

To find out what your route is going to be, enter a five-digit number in your calculator and divide it by another five-digit number (not the same number). The six digits to the right of the decimal point of the result are the ROUTE NUMBER, which represents the cities you must visit. The list below tells you the number of the cities:

1) Los Angeles, California
2) Bismarck, North Dakota
3) Brownsville, Texas
4) Minneapolis, Minnesota
5) Portland, Oregon
6) Denver, Colorado
7) Jacksonville, Florida

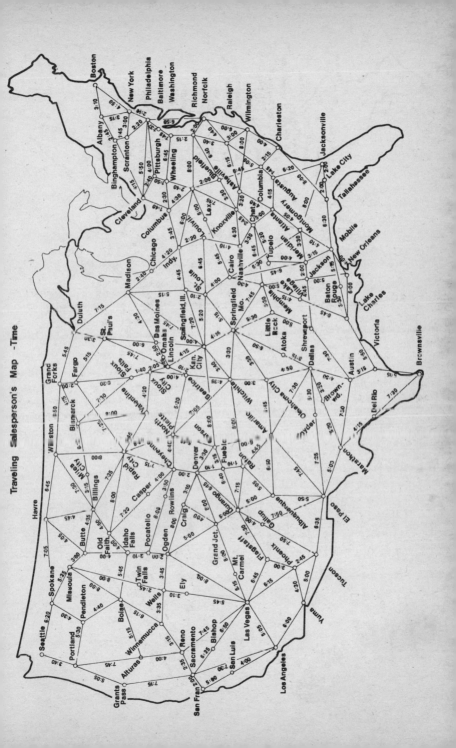

Traveling Salesperson's Map -Time

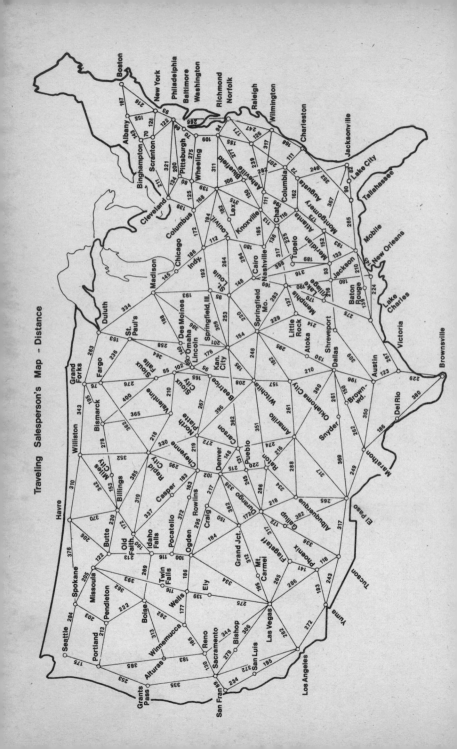

Traveling Salesperson's Map – Distance

8) Pittsburgh, Pennsylvania
9) Chattanooga, Tennessee
0) St. Louis, Missouri

You must visit all the cities in your route number. If you get two of the same digit separated by another digit you must visit that city TWICE. For example, if your route number is .917077, you must visit Jacksonville (7) twice, but not three times. If the same digits are together (77), you only visit that city once. You can end up at any of the cities on your route.

Once you have discovered the cities you must visit, calculate the shortest route by adding the distances you think are the shortest between the cities. To make the fastest trip possible, you must calculate the smallest amount of time for the same trip. (Remember that adding distance is different from adding time. Do not forget to calculate minutes; when they add up to over sixty, change them into hours).

### STRATEGY
Look at the cities that you must visit on your sales trip to try to figure out the shortest route you can take without going over the same roads twice.

### SAMPLE PLAY
Player enters 89654 in calculator, and divides it by 56321 = 1.5918396. Thus, player has to visit:

5 = Portland, Oregon
9 = Chattanooga, Tennessee
1 = Los Angeles, California
8 = Pittsburgh, Pennsylvania
3 = Brownsville, Texas
9 = Chattanooga, Tennessee (again)

The best route seems to be: start in Chattanooga, go to Pittsburgh, then back to Chattanooga, to Brownsville, to Portland, to Los Angeles. What do you think?

# Salesforce

**2 Players**
**2 Calculators**
**Traveling Salesperson's Maps**
**Paper and Pencil**

## OBJECT OF THE GAME

For the two salespersons to work together to sell all their widgets before they spend all their money.

## THE PLAY

Each player starts by entering a five-digit number in his or her calculator, pressing the divide (÷) button, and passing it to the other player. The other player enters a five-digit number and passes the calculator back. The players press the equals (=) buttons. The numbers they get are their ROUTE NUMBERS. All the digits to the right of the decimal points are numbers of the cities they must visit on their trip. Consult the list below and the Traveling Salesperson's Maps.

1) San Francisco, California
2) Billings, Montana
3) Omaha, Nebraska
4) Phoenix, Arizona
5) Indianapolis, Indiana
6) Cleveland, Ohio
7) Duluth, Minnesota
8) Mobile, Alabama
9) Charleston, South Carolina
0) Oklahoma City, Oklahoma

If two digits in a ROUTE NUMBER are the same and adjacent, the player has to visit that city only once.

If the digits are separated, the player must visit that city as many times as its number appears. For example, if the ROUTE NUMBER is .7707893, the player has to visit Duluth (7) twice, but not three times.

The player must start in the city whose number is the first to the right of the decimal point and proceed to the city to the right of that digit, and so on. For example: if the number is .89765434, the first city is Mobile (8), then Charleston (9), and so on, reading the number from left to right.

Players decide who is Player A and who is Player B. Player A pays expenses on the basis of miles. Each 100 miles costs $100. Since the route numbers the players can get may overlap and since the two salespeople CANNOT BOTH SELL WIDGETS IN THE SAME CITY, one or the other of the salespersons may simply pass through the city. This costs them $50. If one stays and sells a widget, this costs them $300. Player A cannot travel more than 480 miles a day and must be in a city every night (at the end of a turn).

Player B pays expenses on the basis of hours. Every two hours cost $100 (omit fractions), and each visit to a city costs $300 if a widget is sold. Passing through the city takes an hour for traffic and parking and so it costs $50. A working day cannot exceed eight hours and forty-five minutes, and Player B must be in a city every night (at the end of a turn).

The two salespeople own their own company. They have seventy-five widgets to sell. They have their own routes (ROUTE NUMBERS). They have only $25,000 in their business and cannot spend more than that on their trip. In this game we do not concern ourselves with how much they get for selling a widget, but if they sell all seventy-five widgets, their company will be a success. If they spend their money before they sell all seventy-five widgets, they will lose their company. Thus, the daily expenses of the two salespeople must be subtracted from the $25,000 total. Every time they visit a city, **THEY SELL A WIDGET**. The two players must work together to sell all the widgets before their $25,000 is gone.

## STRATEGY

The players should compare routes. The only way they can win is if their strategy is cooperative.

## SAMPLE PLAY

Player A enters 45683 ÷; Player B enters 14785 = .08982076 to the right of the decimal point.

Player A starts in Oklahoma City (0) and goes to Mobile (8), to Charleston (9), to Mobile (8), to Billings (2), to Oklahoma City (0), to Duluth (7), to Cleveland (6), **IN THAT ORDER.**

Player B enters 98765 ÷; Player A enters 54789 = .80264286. Player B starts in Mobile (8) and goes to Oklahoma City (0), to Billings (2), to Cleveland (6), to Phoenix (4), to Billings (2), to Mobile (8), to Cleveland (6), **IN THAT ORDER.**

Player A's first day: Starts in Oklahoma City, goes to Dallas, to Austin (210 + 196 = 406). He or she can also pass through one city ($50) 406 + 50 = $456, and sell one widget in one city = $300. Subtract ($456 + 300) $756 from $25,000 = $24,244, and seventy-four widgets remain to be sold.

Player B's first day: Starts in Mobile, goes to Meridian, to Jackson (3:15 + 2:00 = 5:15). Sells three widgets in three cities (this takes three hours) 5:15 + 3:00) = 8:15 costs $400. Three cities cost $900 but Player B sells three widgets. $24,244 − $1300 ($400 + $900) = $22,944 and seventy-one widgets remain to be sold.

Play continues until the two players have sold all the widgets or have lost all their money.

# Widgets

**4 or more Players**
**1 Calculator per Player**
**Traveling Salesperson's Distance Map**
**Paper and Pencil**

## OBJECT OF THE GAME

For all players to sell their seventy-five widgets cooperatively.

## THE PLAY

Each player starts by entering a five-digit number in his or her calculator, pressing divide ($\div$), and passing it to the right. Each player who now holds the calculator enters another five-digit number and passes it back to the left. The players now enter equals ($=$). The numbers on their calculators are their ROUTE NUMBERS. The first six digits to the right of the decimal point represent the cities they must visit. THEY CAN VISIT THEM IN ANY ORDER THEY WANT.

1) Los Angeles, California
2) Omaha, Nebraska
3) Phoenix, Arizona
4) Billings, Montana
5) Cleveland, Ohio
6) Oklahoma City, Oklahoma
7) Charleston, South Carolina
8) Indianapolis, Indiana
9) Mobile, Alabama
0) Duluth, Minnesota

For example, if a player has .987654, he or she must visit Mobile, Indianapolis, Charleston, Oklahoma City, Cleveland, Billings, IN ANY ORDER.

The players are employed by Widget International. They are all trying to sell the company's supply of seventy-five widgets.

There is a fuel shortage. *Players are allotted 150 gallons of gasoline each. The players rent cars that get twenty miles to a gallon.* Therefore, they cannot travel more than 3,000 miles if they travel alone. However, if two players share the same car they get an extra two miles to a gallon. If three players share a car, they get an extra five miles to a gallon. If four players share a car, they get an extra seven and a half miles to a gallon. (This is in addition to the benefit of splitting the gas used two, three, or four ways.)

Since all the players work for the same company, if they are to succeed in selling the seventy-five widgets they cannot sell in the same city. ONLY ONE PLAYER CAN MAKE A SALE IN ANY ONE CITY DURING THE SAME TURN. Players can sell in the same city as long as they are not there during the same turn.

After they have their ROUTE NUMBERS, and have looked at the Traveling Salesperson's Distance Map, the players decide on routes to reach all the cities and sell all the widgets. A salesperson can only go 480 miles per day. *Each 480 mile trip represents that player's turn.* Every player must take a trip of at least 100 miles and not more than 480 miles each turn. Every player must be in a city at the end of a turn. All the players are sharing the responsibility of selling the widgets.

### STRATEGY

Once the routes are determined, try to get as many of the routes as possible to overlap so that gas is conserved.

### SAMPLE PLAY

Player A enters 14253 ÷ and passes it to Player B who enters 58693 =. Player A's ROUTE NUMBER is .242839 = Omaha, Billings, Omaha, Indianapolis, Phoenix, Mobile.

Player B enters 12365 ÷ and Player C enters 14369 =. Player B's ROUTE NUMBER is .860533 = Indianapolis, Oklahoma City, Duluth, Cleveland, Phoenix. (Player B need not visit Phoenix twice, since the two 3's are consecutive.)

Player C enters 58796 ÷ and Player D enters 45632 =. Player C's ROUTE NUMBER is .288481 = Omaha, Indianapolis, Billings, Indianapolis, Los Angeles.

Player D enters 31648 ÷ and Player A enters 32352 =. Player D's ROUTE NUMBER is .978239 = Mobile, Charleston, Indianapolis, Omaha, Phoenix, Mobile.

Next, the players plan their trips on the basis of the cities they must visit on their routes:

Player A and Player B decide to go together from Omaha to North Platte. They will pass through Lincoln and Beatrice.

Player C and Player D decide to go together from Charleston to Columbia, South Carolina, to Augusta, to Atlanta, and to Chattanooga.

The combined route of players A and B was $80 + 95 + 267 = 442$ miles. They visit three cities and each player sells one widget. Since they rode together they get twenty-two miles to the gallon: $442 \div 22 = 20$ gallons (round off to the smaller number). Each player subtracts ten from his or her 150 gallons.

The combined route of players C and D was $111 + 72 + 162 + 116 = 461$ which is fine. They visit four cities and sell two widgets each. They ride together and get twenty-two miles to the gallon, $461 \div 22 = 21$ gallons. The players each subtract ten and a half gallons from their 150 gallons.

In one turn, the four players have sold seven widgets and used up forty-one gallons of gasoline.

# Fast Eddie

**4 Players**
**4 Calculators**
**Traveling Salesperson's Maps**

## OBJECT OF THE GAME

To visit the most cities in the amount of time and distance allotted.

## THE PLAY

All the players must look at the Traveling Salesperson's Maps. Each player has 112 hours and 5,000 miles allotted to him or her, and he must go from the starting city to the finishing city within those limits. Turns are composed of sixteen hours. A player's turn consists of trying to visit as many cities as possible during that sixteen-hour period (adding up the hour value from the map) and adding the distance covered.

Each player chooses one of the starting city/finishing city combinations from the list below. If players want to vary the game, they can select different starting city/finishing city combinations, but the cities must be at least 2,500 miles apart. ANY PLAYER WHO DOES NOT REACH HIS OR HER FINISHING CITY HAS A PENALTY OF LOSING FIVE CITIES FROM THE TOTAL CITIES THAT HAVE BEEN ACCUMULATED THROUGH PLAY.

*Player A:*   starting city is Norfolk, Virginia
               finishing city is Los Angeles, California
*Player B:*   starting city is Jacksonville, Florida
               finishing city is San Francisco, California

*Player C:*  starting city is Pittsburgh, Pennsylvania
              finishing city is Seattle, Washington
*Player D:*  starting city is Spokane, Washington
              finishing city is New Orleans, Louisiana

## STRATEGY

Try to get to cities that have several possible routes of entrance and exit so you can have a choice on your second, third, and later rounds. Be sure that you have enough miles at the end to get to your last city.

## SAMPLE PLAY

Player A starts in Norfolk and goes to Richmond, to Washington, to Baltimore, to Philadelphia, to New York, to Albany, to Binghamton, to Scranton, to Binghamton = ten cities, sixteen hours, and 870 miles.

Player B starts in Jacksonville and goes to Lake City, to Tallahassee, to Pensacola, to Mobile, to New Orleans, to Baton Rouge, to Jackson, Mississippi = eight cities, sixteen hours, and 813 miles.

Player C starts in Pittsburgh and goes to Wheeling, to Columbus, to Indianapolis, to Springfield, Illinois, to St. Louis, to Springfield, Missouri = six cities, fifteen hours and forty-five minutes, and 882 miles.

Player D starts in Spokane, Washington, and goes to Missoula, to Butte, to Old Faithful, to Pocatello, Idaho, to Ogden, Utah = six cities, sixteen hours, and 808 miles.

This completes one turn. Players continue until they have gone 112 hours or 5,000 miles or have reached their finishing city.

## COMPLEX VERSION

The players each have forty cities to visit, and:
Player A starts in New York.
Player B starts in Richmond, Virginia.
Player C starts in Seattle, Washington.
Player D starts in Boise, Idaho.

The players have to try to visit the forty cities in the least number of miles and the fastest time possible. Each turn is ten cities. Each player tries to visit the ten cities in the minimum number of miles and the minimum amount of time in his or her turn. No player can visit the same city more than twice. The players can go anywhere on the map and finish at any city on the map, as long as they have visited forty cities. Each player is allotted 112 hours and 5,000 miles. The starting city counts as one city.

# 49er

**2 Players**
**2 Calculators**
**1 Deck of Cards**

## OBJECT OF THE GAME

To make three sets of cards, each set equaling 49. The game is structured similarly to the card game Gin Rummy (for example: a set would be $10 \times 4 + 9 = 49$).

## THE PLAY

Remove all face cards and jokers from the deck. One player, the dealer, shuffles and deals what is basically a Gin Rummy hand: eleven cards to the opponent and ten cards to the dealer.

To win a hand a player must make three sets of cards; the cards in each set are put together using the operations $+ - \times :$ to make a total of 49. Operations may be used any number of times and are performed serially. Thus, the cards 10, 4, 9, 1 (ace) could be put together by the series $1 \times 10 \times 4 + 9 = 49$. The cards 5, 2, 8, 7 could be used in the series $7 \times 8 - 5 - 2 = 49$.

To begin play, the dealer's opponent arranges his or her cards and discards one card, face up. The dealer may take this card or the top card from the remaining deck. The dealer then discards one card. Play continues in similar fashion until one player achieves a complete hand.

A complete hand is one in which, after discard, the ten cards are part of a series making 49. Since three sets making 49 each are necessary, only certain numbers of cards may be used in each set. Two sets containing four cards each, and one set with three cards would total eleven cards, which is one too many. Sets containing one set of four and two sets of three cards will work, as will combinations of:

$$2 \& 4 \& 4 = 10$$
$$2 \& 5 \& 3 = 10$$
$$2 \& 2 \& 6 = 10$$

When a player makes a perfect hand on a draw, he or she discards the eleventh card and places the cards face up for the other player to check.

The player winning the hand receives twenty points plus the sum of the other player's cards which are not in sets of 49. Thus, if the loser has 10, 3, 5 remaining, the winner receives $20 + 10 + 3 + 5 = 38$ points for the hand. The loser may take two minutes to arrange his cards after a hand is won. Hands are played until one player reaches 100 points.

## STRATEGY

There are many ways to make 49 from cards in a hand. The only way to make 49 with two cards is with two 7's, $7 \times 7 = 49$. With three cards, some combinations are:

$$10 \times 5 - 1 = 49$$
$$8 \times 7 - 7 = 49$$
$$7 \times 6 + 7 = 49$$
$$4 \times 10 + 9 = 49$$

With four cards:

$$4 \times 10 + 6 + 3 = 49$$
$$8 \times 8 - 9 - 6 = 49$$
etc.

It should be obvious that some cards such as 7, 4, and 10 are more valuable than others. In discarding, a player must try to guess what the opponent needs and attempt to discard cards which will not help the opponent make a better hand.

After a hand, the losing player should attempt to use all possible cards in sets of 49 to avoid losing additional points.

# 50

**3 or more Players**
**1 Calculator per Player**
**1 Deck of Cards**

## OBJECT OF THE GAME

To get 50. 50 functions in this game the way 21 does in Blackjack.

## THE PLAY

Remove all face cards and jokers from the deck. The remaining cards are aces (counted as 1) through 10's.

The dealer shuffles and deals (left to right and one at a time) two cards face down to each player and one card face up. Players need not show their down cards.

In the game "21" or Blackjack, players add cards trying to make the sum as close to 21 as possible. In "50" players try to make, but not to exceed, 50. This may be done by combining the cards with any of the functions $(+ - \times \div)$. Thus having been dealt three cards, a player may put them in any order using any two functions. Some examples:

Dealt 5, 10, ace $= 5 \times 10 \times 1 = 50$, a natural
Dealt 7, 7, ace $= 7 \times 7 + 1 = 50$, a natural

(Notice that these cards could have been put in other arrangements, for example $1 + 7 \times 7$, whose total would be $8 \times 7 = 56$.)

In all cases the operations are performed one at a time in the order given.

Dealt 4, 7, 8, then $4 \times 8 + 7 = 39$ is the best you can do.

Additional Cards: If the player is not satisfied with his or her three-card total, the player may ask the dealer for an additional card by saying, "Hit me." This card is dealt face up. *Only one additional card may be dealt.*

For example: Holding 4, 7, 8 before making $4 \times 8 + 7 = 39$, the player calls for another card. The dealer deals a 9. The player can make $4 \times 9 + 7 + 8 = 51$ which is close but too high (a Bust). An option is to make $4 \times 8 + 9 + 7 = 48$ which is close to 50.

Betting: Bets may be placed in many ways. One round of betting may be desired after two face-down cards have been dealt to each player, with another round after the third card is dealt.

Winning: The player whose total is the closest to 50 (without going over) wins all the money. In the game below, Player A wins with 50. If two or more players are tied, they must divide the money equally.

**STRATEGY**

Players should try different arrangements and functions before settling on the best total. For instance, 5, 10, 4, 4 could be played $5 \times 10 \times 4 \div 4 = 50$. In general, one should stand on 48 or better and take a hit on 46 or less. 47 is optional.

**SAMPLE PLAY**

A sample four-handed game might proceed as follows:

Player A dealt 9, 3, 8 making $9 \times 3 + 8 = 35$
Player B dealt 3, 3, 4 making $3 \times 3 \times 4 = 36$
Player C dealt 6, 5, 3 making $3 + 5 \times 6 = 48$
Player D dealt 7, 8, 5 making $5 \times 8 + 7 = 47$

Player A asks for a card, receiving a 7, now making $7 \times 8 + 3 - 9 = 50$.

Player B asks for a card, receiving a 2, now making $3 + 3 \times 4 \times 2 = 48$.

Player C stands on 48.

Player D asks for a card, receiving a 5, now making $5 \times 7 + 8 + 5 = 48$.

A has winning hand.

# Natural Resources

4 Players
4 Calculators
1 Pair of Dice
100 Toothpicks ⎫
100 Matches    ⎪ *or accounting can be done on paper,*
100 Paper Clips ⎬ *without actual items.*
100 Pennies    ⎭
Paper and Pencil

## OBJECT OF THE GAME

To distribute the four hundred items so that the players all have twenty-five of each of the items (or have them accounted for on paper).

## THE PLAY

Each player begins with one hundred of one kind of item. Each kind of item represents a resource. All four hundred represent the total resources of the country. The four players are the representatives of the society and they are trying to distribute the resources so that each representative has an equal amount of the resources. The players can name the resources such as food, metals, autos, houses. The game is played until the representatives all have twenty-five of each resource.

Each player rolls the dice. The highest roll goes first, the next highest second, and so on. The first player rolls the dice again. This number is his or her RESOURCE EXCHANGE NUMBER. The player can use this number in any way desired. He or she can give that number of resources to another player directly, or divide that number into the total resources and give that number away.

On any one turn, a player can only give resources to ONE OTHER PLAYER, and must use the RESOURCE EXCHANGE NUMBER AS THE BASIS OF EXCHANGE. For example: a player may roll 3 and 5. He or she could give three resources to one player and ask for five back of something else. A player can exchange any resources that are in his or her possession at that turn. Thus, a player throws 6, has one hundred resources, and divides 6 into 100: $100 \div 6 = 16.66$ (decimals are rounded off to the lowest number, so $16.66 = 16$). The player then gives sixteen resources to another player.

As resources start to be distributed, it will become more and more challenging to get the RESOURCE EXCHANGE NUMBER to meet the needs of exchange. It is necessary to keep score with pencil and paper so that there is no confusion.

## STRATEGY

Try to distribute small quantities at first so that no player begins to have too many of one resource too early in the game. In this way, the resources can begin to equalize slowly and it will be easier to exchange resources as the players have some of each of them. You can use your RESOURCE EXCHANGE NUMBER AS A PERCENTAGE (8 becomes .08 × resource).

## SAMPLE PLAY

REN = RESOURCE EXCHANGE NUMBER (DICE)

| | Player A | Player B | Player C | Player D |
|---|---|---|---|---|
| START | 100 autos | 100 food | 100 metals | 100 kilowatts |
| ROUND 1: | 8 REN | 10 REN | 11 REN | 2 REN |
| | .08 × 100 = 8 | 100 — 10 = 90 | 100 — 11 = 89 | 100 ÷ 2 = 50 |
| | Gives 8 to B | Gives 10 to C | Gives 11 to B | Gives 50 to C |
| | 92 autos | 90 food | 89 metals | 50 kilo |
| | 0 food | 11 metals | 50 kilo | 0 autos |
| | 0 metals | 8 autos | 10 food | 0 food |
| | 0 kilo | 0 kilo | 0 autos | 0 metals |

| Round 2: | 7 REN | 12 REN | 4 REN | 9 REN |
|---|---|---|---|---|
| | $92 \div 7 = 13$ | $90 - 12 = 78$ | $.04 \times 89 = 3$ | $50 - 9 = 41$ |
| | Gives 13 to D | Gives 12 to A | Gives 3 to D | Gives 9 to A |
| | 79 autos | 78 food | 86 metals | 41 kilo |
| | 12 food | 11 metals | 50 kilo | 13 autos |
| | 0 metals | 8 autos | 10 food | 0 food |
| | 9 kilo | 0 kilo | 0 autos | 3 metals |
| Round 3: | 5 REN | 2 REN | 8 REN | 10 REN |
| | $79 \div 5 = 15$ | $78 \div 2 = 39$ | $86 - 8 = 78$ | $41 - 10 = 31$ |
| | Gives 15 to C | Gives 39 to D | Gives 8 to A | Gives 10 to B |
| | 64 autos | 39 food | 78 metals | 31 kilo |
| | 12 food | 11 metals | 50 kilo | 13 autos |
| | 8 metals | 8 autos | 10 food | 39 food |
| | 9 kilo | 10 kilo | 15 autos | 3 metals |

Play continues until each player has twenty-five of each resource.

**COMPLEX VERSION**

The round (four turns) equals a year. The resources have to be equally distributed within eight years. Every year after eight, the players have to give up five resources to the environment in terms of waste. If the waste gets larger than 50, all the players lose.

# Four Out

**1 Player**
**1 Calculator**

## OBJECT OF THE PUZZLE

To get rid of a six-digit number in four moves so that the calculator will read 0.

## THE PLAY

Select a six-digit number, all digits being different from one another. You now have to get rid of that number in four moves using TWO DIGIT NUMBERS and any operation ($\times$ $+$ $-$ $\div$). Each move consists of using a two-digit number and an operation. You may not multiply or divide by 0.

## STRATEGY

Using divide early in the puzzle may decrease the number rapidly, but you run the risk of developing decimal numbers which are difficult to eliminate. The best strategy is to get the number to a form which is evenly divisible and then divide.

## SAMPLE PLAY

| Six-digit number | 542681 |
|---|---|
| Move 1 | $- 81 = 542600$ |
| 2 | $\div 50 = 10852$ |
| 3 | $\div 52 = 208.6923$ |
| 4 | $\div 99 = 2.10800$ |

The above result is not very good. Move three was crucial and was estimated incorrectly. Once you get the sense of the puzzle, you will be surprised at how fast a number can be eliminated.

## COMPLEX VERSION

Cover up the readout after you have entered the six-digit number and try to reduce the number to zero in as few moves as possible. Do not look at the calculator readout until you think you have reached zero. If you look, the game is over.

If you have a calculator with more complex functions, you can use them to make the game more challenging.

# Covert Operation

**2 Players**
**1 Calculator**

**OBJECT OF THE GAME**

For the two players to cooperate in creating a number using addition and subtraction without looking at the readout of the calculator.

**THE PLAY**

The game is played with one calculator. The players decide on a one-digit STARTING NUMBER and *enter it into the calculator*. They also decide on a two-digit GOAL NUMBER which they will reach by taking turns entering operations and numbers.

The players cover up the readout with a piece of paper, tape or just with a hand. Player A enters an operation (+ or − ONLY) and a number and passes it back to Player B who enters an operation and a number and passes it back to Player A. They continue until they agree that they have reached their GOAL NUMBER. At this point, they uncover the readout and look at it. They have won if it is the GOAL NUMBER. If it is not, they have not won and they try again until they reach the GOAL number exactly.

**STRATEGY**

The players should start the game using low GOAL NUMBERS and low STARTING NUMBERS.

**SAMPLE PLAY**

|  | Initial Number | Player A | Player B | Result |
|---|---|---|---|---|
|  |  |  |  | (GOAL NUMBER = 50) |
| First Round | 5 |  |  |  |
| Move |  | + 20 | + 35 |  |
| Outcome |  | 25* | 60* | 60 (no good) |
|  |  |  |  | (GOAL NUMBER = 80) |
| Second Round | 5 | + 45 | + 30 |  |
| Move |  | 50* | 80* | 80 (goal reached) |

*not seen by players

**COMPLEX VERSION**

Include multiplication and division in the game. The **GOAL NUMBER** should then be five digits.

# Secret Enterprise

**2 Players**
**1 Calculator**
**Paper and Pencil**

## OBJECT OF THE GAME

For the two players to each pick GOAL NUMBERS that they want to reach, and for the players to compete in reaching them.

## THE PLAY

One calculator is used. The players each pick a GOAL NUMBER and write it down. The two players agree on an INITIAL NUMBER and enter it into the calculator. Then the two players take turns entering operations ( |  — × ÷) and numbers until each believes that they have reached their GOAL NUMBER.

After they have entered the INITIAL NUMBER, the calculator readout is covered over with a piece of paper, or tape or by hand. Thus, the first player enters an operation (after the INITIAL NUMBER has been entered) and a number, trying to get his or her GOAL NUMBER, passes it to the other player who enters an operation and a number, and passes it back to the first player.

They continue to pass it back and forth until either one of the players believes his or her number has been reached. They then look at the readout. If it is close to either player's GOAL NUMBER (nearer than 50) that player gets one point. If it isn't close to either player's GOAL NUMBER the players start again.

The game should be repeated several times to familiarize the players with the play. The first player to get eleven points wins.

**STRATEGY**

Each player must try to guess his or her opponent's strategy and stay away from the number that he or she is trying to get to. Dividing and subtracting may be very useful.

**SAMPLE PLAY**

Initial Number: 100; Goal of Player A: 10; Goal of Player B: 1000.

| | Initial Number | Players | | Result |
|---|---|---|---|---|
| | | A | B | |
| 1st Round | 100 | | | |
| Move | | ÷ 100 | × 2000 | |
| Outcome | | 1* | 2000* | 2000 |
| 2nd Round | | | | |
| Move | | ÷ 1000 | × 2000 | |
| Outcome | | 2* | 2000* | 4000 |
| 3rd Round | | | | |
| Move | | × .01 | × 100 | |
| Outcome | | 40* | 4000* | 4000 |
| 4th Round | | | | |
| Move | | ÷ 400 | × 1 | |
| Outcome | | 10 | 10* | 10 |

A wins

*not known by the players

# MindControl

**4 or more Players**
**1 Calculator**

## OBJECT OF THE GAME

The players are divided into two teams. Each team is trying to reach its own GOAL NUMBER first.

## THE PLAY

One player enters, say, 10 in the calculator as the INITIAL NUMBER. The two teams sit so that each player is next to the members of the other team. The first team is trying to get to 100 on the calculator as their GOAL NUMBER, and the second team is trying to get to − 100.

After the INITIAL NUMBER is put in the calculator, the readout is covered up. The first player on Team A enters a number and an operation (+ − × ÷), and passes to a member of Team B who enters a number and an operation, and so on until the calculator is back in the first player's hands. He or she looks at the readout and announces the result. Team A wants the result to be 100, and Team B wants the result to be − 100. The play continues until either 100 or − 100 is reached.

## STRATEGY

Since one team is going for a plus number 100, and the other team is going for a minus number 100, the teams will have to compensate for the values entered by the other team. This may make the numbers very large or very small, but after several rounds the players will begin to know the other team's strategy and act accordingly.

**SAMPLE PLAY**
  INITIAL NUMBER chosen is 10.
Goal of Team A: + 100    Team A: Players A1 and A3
Goal of Team B: − 100    Team B: Players B2 and B4

| | Input | Players A1 | B2 | A3 | B4 | Output |
|---|---|---|---|---|---|---|
| 1st Round | 10 | | | | | |
| Action | | + 100 | — 50 | + 50 | — 100 | |
| Outcome | | 110 | 60 | 110 | 10 | 10 |
| 2nd Round | 10 | | | | | |
| Action | | + 200 | — 100 | + 80 | — 100 | |
| Outcome | | 210 | 110 | 190 | 90 | 90 |
| 3rd Round | 10 | | | | | |
| Action | | + 300 | — 200 | + 50 | — 260 | |
| Outcome | | 310 | 110 | 160 | — 100 | — 100 |

Team B wins

**COMPLEX VERSION**
  One calculator is still used in this cooperative version. Player A enters, say, 10 in the calculator, and states this to the other players. A covers up the readout with a piece of paper, tape, or with his or her hand. He or she then enters an operation (+ − × ÷), and another number, and passes it to the right. Player B enters an operation, another number, and so on, until the calculator is passed back to Player A.
  Player A reads the readout and announces the number. The other players do not see the readout or know the entries of the other players until Player A gets it back again and announces the result.
  The aim is, of course, to get to 100. The game must be played many times so that players begin to get a mutual sense of their play and their cooperation will lead to the goal number of 100.

# Economy

**4 Players**
**4 Calculators**
**500 Chips**
*(or accounting can be done on paper)*
**1 Pair of Dice**
**Paper and Pencil**

## OBJECT OF THE GAME

This is a game for four role players: General Industry, Ralph Consumer, Sam Government, and David Bank. The players are each trying to get the most chips (or toothpicks) with the least expenditure of their own resources.

## THE PLAY

The players each are allotted 100 chips:

General Industry: each chip is worth 100 industrial units (IU)

Ralph Consumer: each chip is worth 100 hours (H)

Sam Government: each chip is worth 100 promises (P)

David Bank: each chip is worth 100 dollars (D)

The play of the game involves the operations of a small economic society. After every four turns (one turn per player), it is a new year. Ralph Consumer gets twenty-five more chips every year (25 H), since each year there are more hours to work. The other players can only negotiate to increase their wealth. The basic exchange rate is 100 HOURS = 100 INDUSTRIAL UNITS (1 H = 1 IU).

All the other rates, bank rates and promise exchange rates are variable. The PROMISES are cashed in after four turns (at the end of every year).

David Bank decides on lending and borrowing rates, and negotiates with the other players for money and other considerations. However, the interest rate can only be between 1 percent and 20 percent. The players are each trying to make the most of his or her resources and not spend any money. So, the players are trying to get money chips for the fewest number of their units, H, P, or IU.

The turn of a player represents his or her attempt to optimize the situation in his or her terms. Since Ralph Consumer cannot use General Industry's units and can use Dollars, the trading between these two players cannot be direct. In fact, no trading can be direct. ALL TRADES MUST GO THROUGH THE BANK. The players can make several trades in a turn always figuring in the percentage rates that the Bank is charging and other factors. In any turn, a player is trying to get rid of some of its units in exchange for Dollars, and other units which he or she thinks will be more secure. However, no player may TRADE AWAY ALL RESOURCES IN ONE TURN.

For example, Ralph Consumer may buy Industrial Units and Promises in one turn. General Industry may sell Industrial Units for money to the bank. Or any player may buy PROMISES from Sam Government. Promises are the way that the deals of a turn can be assessed. (See Promises Roll below.) After four turns, the players each roll the dice. This is the PROMISES ROLL. To protect against the PROMISES ROLL, a player may put his or her money in David Bank. This may or may not protect it completely, but at least the player does not make a PROMISES ROLL. If a player has no money, no D chips, THE PLAYER DOES NOT MAKE A PROMISES ROLL. The bank always makes a PROMISES ROLL.

THE PROMISES ROLL: Each player who has money (D) throws the dice. If a player rolls an *even number*, this represents the INFLATION INDEX.

The player multiplies the dice number by the number of PROMISES that he or she has. The INFLATION INDEX is a percentage, so if the player rolls 12, for instance, he or she multiplies .12 by the number of promises, and then subtracts this from the number of money chips (D) the player has.

For example: if a player has twenty-five promises (P), and thirty-five money chips (D), and rolls 8, he or she does the following: $25 \times .08 = 2, 35 - 2 = 33$. The player gives two D to the bank.

If a player rolls an *odd number,* this represents the INTEREST RATE. To calculate the interest rate, the player multiplies the number by the number of PROMISES he or she has, and the result of this the player receives from the BANK. For example: the player rolls 11 and has twenty-five PROMISES: $25 \times .11 = 2.75$. Round this off to 3 (all decimal numbers round off to the next HIGHEST number). The player gets three D from the BANK.

Any player having money must do the PROMISES ROLL. Sam Government must also do the PROMISES ROLL. The PROMISES ROLL of David Bank may involve all the players since they may have their money in the Bank. The Bank rolls the dice. If it is even, the Bank multiplies this number times the number of promises they have, subtracts this from the total DOLLARS, and gives it to Sam Government. If the Bank rolls an odd number, the Bank multiplies this times the number of promises it has, and asks the Government to GIVE that number of D to the Bank as Interest from the TREASURY. The first player to have more than 350 chips at any one time wins.

## STRATEGY

Each player has a different strategy. The game has to be played with a sense of the reality of the situation. General Industry wants to get many HOURS, many PROMISES, much MONEY, and produce few IN-

DUSTRIAL UNITS. Ralph Consumer wants all the MONEY, and does not want to give up any HOURS, and wants all the PROMISES. Sam Government wants the MONEY, the most INDUSTRIAL UNITS, and does not want to give up any PROMISES. David Bank wants all the MONEY.

## SAMPLE PLAY

| | DAVID BANK | RALPH CONSUMER | GENERAL INDUSTRY | SAM GOVERNMENT |
|---|---|---|---|---|
| | 100 Chips (D) | 100 Chips (H) | 100 Chips (IU) | 100 Chips (P) |
| PLAY 1: | 75 H for 63 D | Trades 75 H to Bank for 63 D | 45 IU for 45 D | 18 P for 22 D |
| | 20 IU for 18 D | 15% interest | Trades 20 IU for 18 D (10% interest) | Trades 50 P for 15 D to Bank. Buys 20 IU from Bank for 20 D |
| | 50 P for 15 D | Buys 45 IU for 45 D | Buys 18 Promises for 22 D | 10 P for 10 D |
| | Gets 20 D | 10 P for 10 D | Has 35 IU, 41 D | Puts 25 D in Bank |
| | 6 D (savings) | Has 8 D, 25 H | Puts 30 D in Bank | Has 2 D |
| | 30 D | Puts 6 D in Bank | Has 18 P | 22 P |
| | 25 D | Has 45 IU | 35 IU | 20 IU |
| | Had 4 D | 25 H | 11 D | |
| | | 2 D | | |
| | Has 85 D | 10 P | | |
| | 75 H | | | |
| | 50 P | | | |
| PROMISES ROLL: | Rolls 8 even | Rolls 9 odd | Rolls 6 even | Rolls 12 even |
| | INFLATION | INTEREST | INFLATION | INFLATION |
| | .08 × 50 P = 4 | .09 × 10 = 1 | .06 × 18 P = 2 | .12 × 22 = 3 |
| | Give 4 D to Bank | Gets 1 D from Bank | Gives 2 D to Bank | Gives 3 D to Bank |

Play continues in this manner until one player has more than 350 Chips. *Remember Ralph Consumer gets twenty-five more chips (H) after four turns.*

## COMPLEX VERSION

A fifth player can be introduced into the game. Louise Lobby can take a percentage of every deal between any of the players and do all the negotiating.

# Calcumaze

**1 Player**
**1 Calculator**

### OBJECT OF THE PUZZLE

To find the lowest possible path (number) through the Calcumaze. This path is arrived at by multiplying each length of the maze by the next length.

### THE PLAY

Each branch of the maze is associated with a number. Think of this number as representing the time it takes to walk across that branch. To form a path, multiply succeeding length numbers together.

In the following example, the top path has a length of $2 \times 1.5 \times 1.8 = 5.4$.

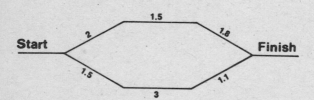

The bottom path has a length of $1.5 \times 3 \times 1.1 = 4.95$. The bottom path, represented by a smaller number, takes a shorter time to travel. Thus, the "length" of the path is the total number for the path and not the number of branches of which it is comprised. One path may have eight branches and still be more optimal than a path containing only four branches.

The only restrictions on a path are that it be continuous and connected at all points, and that it does not use the same branch twice. The same branch may be used in different paths.

The optimal path for the following maze is less than 3.0 (the exact length of the shortest path is 2.9309952).

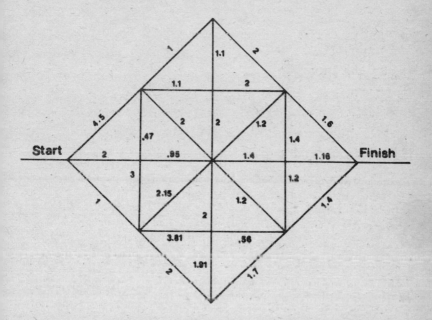

# Lover's Maze

**2 Players**
**2 Calculators**

## OBJECT OF THE GAME

The players each try to find the shortest path (number) through "The Lover's Maze." This path is arrived at by multiplying each length by the next length.

## THE PLAY

One player uses the circled numbers for multiplication and the other player uses uncircled numbers. They take paths together, each calculating part of the path number. The optimal path will have the smallest number when the two players add their path totals.

In the example below, there are three paths. In the

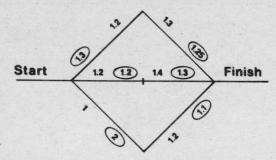

bottom path, Player A (uncircled) has a path $1 \times 1.2 =$ 1.2, and Player B (circled) has a path $2 \times 1.1 = 2.2$. In the middle path, Player A has a path $1.2 \times 1.4 = 1.68$, and B has a path length $1.3 \times 1.2 = 1.56$. In the top path, Player A has a path $1.2 \times 1.3 = 1.56$, while Player B has a path length $1.3 \times 1.25 = 1.625$.

Note that Player A's shortest path is the bottom path and Player B's shortest path is the middle path. However, if Player A's path is combined with Player B's in each case:

| | | |
|---|---|---|
| *Bottom* | A + B = 1.2 + 2.2 | = 3.4 |
| *Middle* | A + B = 1.68 + 1.56 | = 3.24 |
| *Top* | A + B = 1.56 + 1.625 | = 3.185 |

The top path is optimal.

If the players work together, they can test for the optimal path in the following maze. The optimal path has a combined "length" of less than 10 (9.92905056).

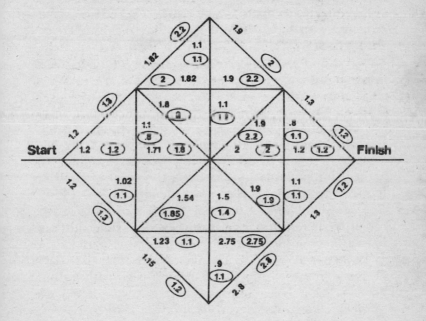

# Dice Maze

**2 or more Players**
**2 Calculators** -
**1 Pair of Dice**
**Paper and Pencil**

## OBJECT OF THE GAME

The players compete to go through "The Dice Maze" from start to finish before the other players. To get through the maze, they multiply successive branches to arrive at a total number equal to the number rolled on their dice.

## THE PLAY

The branches of "The Dice Maze" each have a number which represents its "length." Starting at the left, a path is made by multiplying branch numbers together. For example: the path which goes directly from the start to the finish has a total length of $1 \times 1.25 \times 1.33 \times .8 \times 1.25 \times 2.5 \times 1.1 \times 1.2 = 5.48625$. Paths can cross themselves or go in reverse directions, but one branch may not be used twice in the same path.

To begin a round, each player rolls one die. Each player attempts to run the maze in a path with a number total that is as close as possible to his or her die number. After everyone rolls, players take three minutes to decide on a path and make note of it.

At the end of the decision period, players specify their paths and calculate the "length." They subtract the die number rolled from this length. The result is their score for the round (disregard any negative signs). The first player to reach three or more loses.

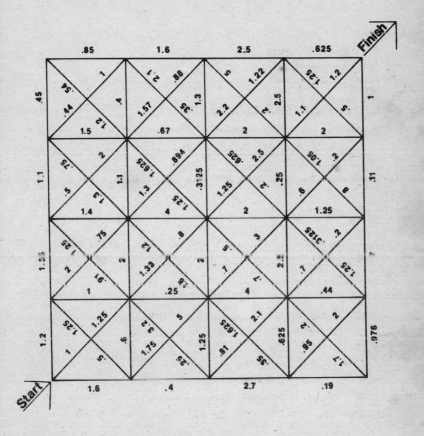

# Commander-in-Chief

1 Player
1 Calculator
1 Pair of Dice
Paper and Pencil

## OBJECT OF THE PUZZLE

To decide how many nuclear arms your country builds and whether or not your country is at war with another nation. You must try to keep the country out of danger for the next ten years.

## THE PLAY

The first year you may choose to have from zero to one hundred nuclear weapons. On each succeeding year, you may increase this amount by a maximum of one hundred weapons.

Each year there is a possibility of war with another nuclear power. Your enemy may also have between zero and one hundred weapons. To determine whether you are at war, roll a pair of dice. If the dice show 5, 6, 7, or 8, you are at war. If you roll any other number, there is no war and you can adjust your weapons number.

If you are at war, you must estimate the number of weapons your enemy has. You know the number of your weapons, which is the sum of your yearly increases and your initial amount. Knowing this, you write a guess of the enemy's arms.

To find out the number of arms your enemy has, roll the dice again. The number showing on the dice gives the number of weapons using the following formula: Dice number × year number × 10. Thus if the dice show 1 and 5 in the third year, the number of weapons is $1 + 5 = 6$, $6 \times 3 \times 10 = 180$. If the dice show 11 or 12, roll them again.

If you have more weapons than the enemy number, you will win the war.

In either case, win or lose, you must calculate to see if there was a nuclear holocaust. To do this, subtract your guess of the other nation's weapons from the actual number (you guess 140 and they have 180 resulting in a difference of $140 - 180 = -40$). Multiply this number (disregarding negative signs) by the number of your weapons and multiply again by the actual number of the other country's weapons. This gives the holocaust number: (Your Guess − Their Weapons) × Your Weapons × Their Weapons = Holocaust number.

If the holocaust number is 1,000,000 or greater, you, the country, and the world are wiped out.

Otherwise, play continues to the next year. If you have lost the war, and there is no nuclear holocaust, multiply the number of your weapons by .25 and subtract that number from the number of your weapons (you lose 25 percent of your weapons). If you had 170 weapons, you would lose $170 \times .25 = 42.5$, and have left $170 - 42.5 = 127.5 = 127$ (rounded).

Each year is played in turn from one to ten, or until there is a holocaust. If you can make it through ten years, you win the Nobel Peace Prize.

# Cold War

**2 Players**
**2 Calculators**
**1 Pair of Dice**
**400 Chips** *(or accounting can be done on paper)*

## OBJECT OF THE GAME

To simulate the way that international affairs are carried out. The players are trying to remain powerful, but not so much that it causes the total destruction of the world.

## THE PLAY

The players each represent a nuclear power (United States, USSR, et cetera). The turns represent a year in the history of the world, starting in the present. The players have one hundred chips each to represent resources of their country. The players each decide upon a WEAPONS NUMBER, which can be any number between 1 and 100 and enter it in their calculator: after each turn, players may add to their weapons number by any number between 0 and 100. (No player can have less than one weapon.) Each turn is directed by one of the players. The players take turns being the DECISION MAKER. A turn represents a decision. A DECISION MAKER can decide one of four possible conditions.

1. To Declare a Peace Conference: The player then rolls the dice and the other players roll the dice. The players must reduce their weapons number by the percentage of the number on their dice (for example: if a player rolls 6, that player reduces the weapons

number by 6 percent. To do this, take the weapons number and multiply it by .06 and subtract the result from the weapons number).

2. TO DECLARE A TRADE AGREEMENT: The players each roll the dice and take the percentage number on the dice, subtract it from the resources, and give it to the other player (for example: if Player A has one hundred resources and rolls 8, then .08 × 100 = 8. Player A gives B eight resources).

3. TO DECLARE WAR: To declare war, the player states "I declare war and I estimate that our weapons numbers combined equal a certain number." The other player can say, "Yes, but I think the number is this," or "No, I don't think we have so many weapons." If the answer is Yes, the first player can make another estimate. This continues until the answer is No, in which case the players must reveal their weapons numbers. If the last estimate made is equal to or larger than the total weapons of the two players, the war is won by the player who declared war. If the combined weapons number is less than the estimate, no one wins the war and each player must subtract 25 percent of the weapons number.

If the war is won, there is the BATTLE. The BATTLE is fought by the player who won. The player adds his or her weapons number to the other player's weapons number and subtracts the estimate which started the war. Then he or she multiplies this times each of the weapons numbers. If the result does not exceed 1,000,000, the loser pays the winner a DEFEATED PENALTY. The loser throws the dice and gives the winner that percentage number of his or her resources.

If the result of the battle exceeds 1,000,000, there is NUCLEAR HOLOCAUST. The players and their resources are now worthless and the game is over.

4. To Pass:   This means that nothing happened that year.

A player can declare war at any time, even if it is not his or her turn. War can break out any time, but players should exercise caution in order to avoid creating NUCLEAR HOLOCAUST.

The play continues until one player has more than 250 resources or until there is nuclear holocaust or until there is disarmament.

Summary of play:
1. Pick weapons number.
2. Decision maker chooses peace, trade, war, or nothing.
3. One of these is carried out:
   a. To declare peace:
      1. Roll dice.
      2. Subtract percentage number of dice from weapons total.
      3. Prepare for next year. (Add or subtract to weapons number.)
   b. To declare trade:
      1. Roll dice.
      2. Exchange percentage of resources shown on dice.
      3. Prepare for next year. (Add or subtract to weapons number.)
   c. To declare war:
      1. Declare estimated weapons number.
      2. Respond until "no" is stated.
      3. Show weapons number.
      4. If attacker is equal or higher
         a. Attacker adds both weapons numbers together and subtracts estimate.
         b. Loser multiplies this number times both original weapons numbers.
         c. If this is greater than 1,000,000, NUCLEAR HOLOCAUST occurs and game is over.

    d. If this is less then DEFEATED penalty must be paid.

    e. Loser throws dice and gives this percentage number to attacker.

    f. Preparation for next year begins. (Add or subtract to weapons number.)

  5. If attacker's weapon number is lower, both players subtract 25 percent of weapons from weapons number and no one wins the war and play continues.

  d. To declare a pass and nothing happens.

4. Next year another player is decision maker.

5. Play continues until complete disarmament or until NUCLEAR HOLOCAUST, or until one player has more than 350 resources.

**SAMPLE PLAY**  (S = Sum of Weapons; LE = Last Estimate; HN = Holocaust Number)

| YEAR | PLAYER A WEAPONS NUMBER | PLAYER B WEAPONS NUMBER | PLAY | CALCULATIONS |
|------|------|------|------|------|
| 1975 | 50 | 45 | Peace | A dice = 6  50 × .06 = 3<br>50 — 3 = 47 new weapons number<br>B dice = 8  45 × .08 = 3.6<br>45 — 3 = 42 new weapons number |
| 1976 | 47 + 28 = 75 | 42 + 28 = 70 | Trade | B dice = 3, has 100 resources<br>100 × .03 = 3, gives to A<br>A dice = 5, has 100 resources<br>100 × .05 = 5, gives to B |
| 1977 | 100 | 100 | Pass | |
| 1978 | 180 | 190 | Pass | |
| 1979 | 150 | 270 | War | B guessed 400 correctly and wins war. A said there were less weapons.<br>S = 150 + 270 = 420; LE = 400;<br>S — LE = 420 — 400 = 20<br>Product: 150 × 270 × 20 = 810,000<br>(less than HN) |
| 1980 | 150 | 270 | Pass | |
| 1981 | 250 | 270 | War | B guessed 500 correctly and has apparently won the war, but<br>S = 250 + 270 = 520; LE = 500;<br>S — LE = 520 — 500 = 20;<br>Product: 250 × 270 × 20 =<br>1,350,000 HN |

NUCLEAR  HOLOCAUST

# Détente

4 **Players**
4 **Calculators**
1 **Pair of Dice**
400 **Chips** *(or accounting can be done on paper)*

## OBJECT OF THE GAME

To carry on world political activity without risking the loss of everything.

## THE PLAY

The players each represent a Nuclear Power (United States, USSR, et cetera). This game is similar in play to COLD WAR except that players can now form alliances, fight wars together, or reduce arms together, et cetera. Players can honor treaties or not, depending on their perspective. Players can say they will reduce arms and not do it, or they can say they will never attack and then attack—just as in the real world. As in COLD WAR, players can increase or decrease their weapons each year by any amount between 0 and 100.

The players will each take turns being the DECISION MAKER. The first Decision Maker will be decided by the highest roll of the dice. The next is the player to his or her right, and so on. Turns are a year. The player who is Decision Maker declares:

1. A PEACE CONFERENCE: Only those nations who want to attend are obliged to. Those nations that attend roll the dice and subtract the number percentage from their total weapons number. If no one decides to attend the Peace Conference, the Decision Maker who

declared it must still roll the dice and subtract the percentage from his or her weapons number.

2. A Trade Agreement: Only those nations that want to attend will attend. The players who attend will roll the dice, subtract the percentage number from their resources, and put that subtracted portion into the center. The attending players roll the dice again and the highest roll takes all the resources that have been put into the center. If no one attends the trade agreement, the Decision Maker who declared it must roll the dice, subtract the percentage from his resources, and divide this equally among the other players.

3. Nothing Is Declared: The next player becomes the Decision Maker. At any time, even if a player is not the DECISION MAKER, he can declare war.

4. War: War can be declared by one player to another player. but before the calculations are made the alliances must be clarified. If a nation goes to war, it may ask another nation to join it. If another nation or other nations agree, this cannot be reversed until the war is over. The player who declared war will make the estimates and decisions and only that player can accept spoken advice. The players cannot move from their places during a war and cannot show their calculations to one another. The player who declares war does not see the weapons of his or her allies until after the estimates are made. (Only when the players are calculating to see if there has been a Nuclear Holocaust do the allied nations contribute their weapons number to the player who has declared war.)

The player who attacks (ATTACKER) makes an estimate of the combined weapons number of the player he or she attacks (VICTIM). The VICTIM can either say that the estimate is "good," in which case the AT-

TACKER must make another estimate, or the VIC-
TIM can say it is "bad." If it is bad, then the players
involved in the war show their weapons number. If
the ATTACKER'S (plus any allies') weapons number
is greater or equal to the VICTIM'S weapons number,
the ATTACKER wins the war, and the VICTIM (plus
any player in the VICTIM'S alliances) rolls the dice
and gives the ATTACKER that percentage of their
resources. They then BATTLE.

To BATTLE, the ATTACKER subtracts his or her
estimate from the total of the weapons of the VICTIM
(or allied victims) plus the ATTACKER'S weapons
number and multiplies this times all the weapons numbers. If this exceeds 1,000,000, there is NUCLEAR
HOLOCAUST and the game is over.

If the ATTACKER'S estimate of the total weapons
is not greater than the VICTIM'S, the players ALL
reduce their weapons number by 25 percent and the
game continues.

The player who continues to play and has more than
three hundred resources will win the game if there is
no NUCLEAR HOLOCAUST or no disarmament.

*Summary of Play:*
1. Pick weapons number and roll for first Decision
   Maker.
2. Make alliances, treaties, et cetera.
3. Decision Maker chooses either

A. Peace Conference: Each player who chooses to
attend rolls dice. If no players attend, Decision Maker
still reduces weapons. Each player subtracts percentage
of weapons that dice show.

B. Trade Agreement: All the players attend who
want to. Players roll dice and put percentage in the
middle. All players roll dice again and highest roll gets
resources in the center. If no one but DECISION

MAKER attends, he or she rolls dice and gives that percentage of his or her resources to each player.

C. NOTHING:   Nothing happens and next turn begins.

D. WAR:   ATTACKER makes estimate of total weapons. VICTIM can agree or disagree with estimate by stating "good" or "bad." This process continues until Victim(s) states "bad," and then all players in war show weapons numbers. If Attacker number is higher and estimate is greater than weapons number total of all players involved, the Attacker wins and the Victims roll dice and give Attacker that percentage of their resources.

If Attacker's weapons number and estimate are smaller than victims, all players in war reduce weapons number by 25 percent. If Attacker was larger, he or she adds up all weapons numbers of that war, and, if there are allies, subtracts his or her estimate from this number and multiplies this times each of the weapons numbers involved in the war. If this is less than 1,000,000, play continues. If it is equal to or greater than 1,000,000, there is NUCLEAR HOLOCAUST and the game is over.

### SAMPLE PLAY

(WN = Weapons Number; RN = Resources Number;
PC = Peace Conference; TA = Trade Agreement)

| | | A | | B | | C | | D | |
|---|---|---|---|---|---|---|---|---|---|
| *Calculations* | *Year Play* | WN | RN | WN | RN | WN | RN | WN | RN |
| | 1975 No | 10 | 100 | 50 | 100 | 50 | 100 | 20 | 160 |

Peace conference all
attend
A rolls 6: 10 × .06
   = 0 subtract none

| Calculations | Year Play | A WN | A RN | B WN | B RN | C WN | C RN | D WN | D RN |
|---|---|---|---|---|---|---|---|---|---|
| | 1975 No | 10 | 100 | 50 | 100 | 50 | 100 | 20 | 160 |

B rolls 4: 50 × .04
= 2 subtract 2
C rolls 1: 30 × .01
= 0 subtract none
D rolls 10: 20 × .10
= 2 subtract 2

| | 1976 PC | 10 | 100 | 48 | 100 | 30 | 100 | 18 | 100 |

B rolls 6: 100 × .06
= 6
C rolls 8: 100 × .08
= 8
D rolls 5: 100 × .05
= 5
A rolls 8: 100 × .08
= 8

A rolls 5, B rolls 6
C rolls 11, D rolls 9,
therefore C has
highest number and
C gets the bundle

| | 1977 TA | 25 | 92 | 65 | 94 | 50 | 119 | 20 | 95 |

C guesses correctly
130 and A rejects. C
wins the war be-
cause 60 + 20 is
larger than 60 so
summing for battle
60 + 20 + 60 =
140. Estimate was
130, 140 — 130 =
10. 60 × 60 × 20
× 10 = 720,000 no
holocaust and game
continues after de-
feated penalty paid.
A rolls 8
  92 × .08 = 7
  92 — 7 = 85
C gets 7 and
  gives 4, 2

| | 1978 War | 60 | 92 | 95 | 94 | 60 | 119 | 20 | 95 |

(Player C declares war on Player A and
makes an Alliance with Player D.)

| | 1979 No | 60 | 85 | 95 | 94 | 50 | 124 | 22 | 97 |

# The Dating Game

**2 Players**
**2 Calculators**

## OBJECT OF THE GAME

For both players to learn more about each other by trying to arrive at the same number.

## THE PLAY

The players enter a five-digit number in their calculators. They do not tell each other what this number is. In fact, other than the prescribed interaction, they may not talk at all. The players try to arrive at the same number.

To communicate, the first player states either:

(1) "I think your number is larger than mine so give me such-and-such a number."

(2) "I think my number is larger than yours so I will give you such-and-such a number."

If Player A states (1), Player B subtracts the amount stated and Player A adds it. If Player A says (2), he or she subtracts that amount and Player B adds it to his or her amount.

The play continues until both players feel that they have the same number. Only then can they speak and show their calculators to each other.

The game should be played many times so that players begin to develop a way of signaling to one another through their number requests.

## STRATEGY

Methods of signaling the other player about your number: you can ask for a large number indicating

132

that you have a small number, or for a small number if you have a large number. Obviously, if both players achieve the same score, they both win.

## SAMPLE PLAY

| STARTING NUMBER: | PLAYER A: 12345 | PLAYER B: 54321 |
|---|---|---|
| PLAYER A: Give me 500 | 12345 + 500 = 12845 | 54321 − 500 = 53821 |
| PLAYER B: I'll give you 1000 | 12845 + 1000 = 13845 | 53821 − 1000 = 52821 |
| PLAYER A: Give me 20000 | 13845 + 20000 = 33845 | 52821 − 20000 = 32821 |
| PLAYER B: Give me 1000 | 33845 − 1000 = 32845 | 32821 + 1000 = 33821 |
| PLAYER A: Give me 500 | 32845 + 500 = 33345 | 33821 − 500 = 33321 |

B states: I think we are close. The players show their calculators to each other. They are 24 apart. Not bad for a first date.

## COMPLEX VERSION (THE MARRIAGE GAME)

Each player enters a four-digit number in his or her calculator. They do not tell their numbers to each other. NO TALKING except for play statements is allowed. There are five play statements by which the players attempt to get to the same number:

1. I think your number is larger than mine, so you subtract _____ .

2. I think your number is smaller than mine, so I will subtract _____ from mine and you will add it to your number.

3. I think your number is much larger than mine, so you divide your number by _____ and I will multiply mine by the same number.

4. I think my number is much larger than yours, so I will divide my number by _____ and you multiply your number by the same number.

5. I think we have the same number, so let's look.

(Decimal numbers are to be disregarded and discarded after each turn.)

If either player gets a number which exceeds the capability of the readout, the game is over and no one wins. The players must use play statements 3 or 4 at least once every three turns.

The two players are trying to get to a point where they feel that they both have a similar or exactly the same number. The quicker they learn the strategy and way of playing, the faster they will be able to get to the same number. Each game is scored by the difference between the numbers. If a game total exceeds 999,999, the game is over. The players can use any numbers in the play statements.

# Calcumeasure

**1 Player**
**1 Calculator**

## OBJECT OF THE PUZZLE
To test the skill of your calculator.

## THE PLAY
Listed below are several series of simple calculations. The series are designed to begin with a certain number and end with the same number. The problem is that your calculator may not be as smart as you think.

The difficulty of these series (for your calculator) progressively increases. Test your calculator by seeing how close it will come to the correct answer.

First try this series:

$$2 \times 4 + 9 \times 6 + 8 \div 10 - 3 \div 4 = 2$$

Any four-function calculator should do this with no difficulty. If your calculator failed this test consider buying a better model.

The next series is a bit tougher:

$$48 \div 1.5 \times 1.75 + 9.5 \times 2 \times 1.1 - 0.1 \div 3 = 48$$

Most calculators should handle the fractions with no difficulty.

This one should eliminate a few:

$$100 \div 5000 + 1 + 100 \times 2000 \times 12 \div 16000 - 51.53 = 100$$

If your readout has only six digits, it may have gotten stuck.

The series below will test the precision of your calculator:

$$3.1415926 \times 24 \div 3 \times 7 \times 56 \div 64 \div 49 = 3.1415926$$

How many digits were wrong?

The next set of four is the crucial test. If your calculator passes, give it a pat on the back.

$$4 \div 5 \times 5 = 4$$
$$4 \div 3 \times 3 = 4$$
$$8 \div 7 \times 7 = 8$$
$$1 \div 9 \times 9 = 1$$

If your machine failed this test, don't feel too bad. Many very expensive machines will not pass. The machines that pass are able to round off the last digit in a calculation.

Try some really big numbers:

$$2 \times 900000 \times 10000 \div 9000000 \div 1000 = 2$$

Unless your calcuator has scientific notation it probably gave you the error or overflow sign.

Finally, to test the logic of your calculator, try the following calculations.

Sometimes numbers and operations are grouped together for easy calculation, such as:

$(2 \times 3) + (4 \times 5) = ?$ which means multiply 2 by 3 and 4 by 5 and add the two results, thus $2 \times 3 = 6$, $4 \times 5 = 20$, and $6 + 20 = 26$.

Now try performing the calculation in a direct manner: $2 \times 3 + 4 \times 5 = 26$, or did you get 50?

If you got 50, your calculator has serial logic, performing all operations one after another. If you got 26, your calculator has sum of products logic, doing multiplication and division first and addition and subtraction last.

# Series Solitaire

**1 Player**
**1 Calculator**
**1 Deck of Cards**

**OBJECT OF THE GAME**

To remove all the cards by discovering a way to add, subtract, multiply, or divide adjacent cards.

**THE PLAY**

Remove all face cards and jokers from a deck of cards. Shuffle the cards and place them face up in a line, one card next to the other. Now, looking at the cards, see if you can find a situation in which two (or more) cards lying next to each other add up to the card to the right.

```
+-------+       +-------+       +-------+
| 3     |       | 6     |       | 9     |
|       |  +    |       |  =    |       |
|     3 |       |     9 |       |     6 |
+-------+       +-------+       +-------+
```

Or see if you can find a set in which two cards lying next to each other if subtracted from each other equal the number of the card to the right.

```
+-------+       +-------+       +-------+
| 10    |       | 7     |       | 3     |
|       |  -    |       |  =    |       |
|    10 |       |     7 |       |     3 |
+-------+       +-------+       +-------+
```

Or see if you can find a set in which two cards lying next to each other when multiplied equal the number of the card to the right.

$$ \boxed{2 \quad 2} \times \boxed{4 \quad 4} = \boxed{8 \quad 8} $$

Or divided.

$$ \boxed{8 \quad 8} \div \boxed{2 \quad 2} = \boxed{4 \quad 4} $$

This is how to play the game. You want to remove all the cards by adding, subtracting, multiplying and dividing cards. When a card is removed, the cards that were on either side of it are now considered adjacent. That is, if you take three cards out, the cards to the left and right of those three cards are now considered next to one another.

If all the cards are removed in this way, the game is won. If any cards remain the game is lost. Remember, any number of cards may be combined.

$$ \boxed{7 \quad 7} + \boxed{3 \quad 3} - \boxed{2 \quad 2} \times \boxed{A \quad A} = \boxed{8 \quad 8} $$

**STRATEGY**

In removing cards with combination, subsequent moves should be considered. This is especially critical near the end of the game. Do not remove cards so that only one or two remain (in which case you lose). Also try not to remove cards so that no possible combinations remain. The length of combinations which can be removed at one time is limited only by your imagination and skill. Could you remove all forty at once?

## COMPLEX VERSION

A variation which makes the game more difficult can be played by using the complete pack of fifty-two cards, dealing all cards as above. Play is the same as above with Jacks counting 11, Queens 12, Kings 13. All other rules are the same.

# Double Solitaire

2 Players
2 Calculators
1 Deck of Cards

**OBJECT OF THE GAME**

To remove all the cards before your opponent does by adding, subtracting, multiplying, or dividing adjacent cards.

**THE PLAY**

Remove the face cards and jokers from a deck of cards. Players lay the cards face up in a line between them, after having taken one card from the deck at random and discarded it. The play is the same as Series Solitaire. The player makes an addition, subtraction, multiplication, or division with adjacent cards.

For example:

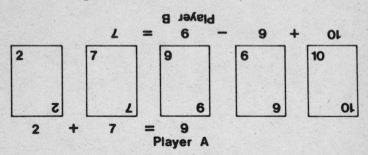

Play begins when all cards have been dealt and continues until neither player can make a series with the remaining cards. Play is simultaneous. The players use their calculators to verify a series to each other, before removing cards. The player with the most cards at the end of the game wins.

Note that one player may be attempting to construct a series only to have the other player remove some of the cards by making a short series on his or her side. In the example give above, if Player B is trying to match up 10, 6, 9, 7, and Player A declares $2 + 7 = 9$, Player B will not be able to make his or her series because 7 and 9 have been removed.

## STRATEGY

When play begins, the players should try to remove as many "easy" series as possible ("easy" could mean three-card sets). After these have been exhausted, longer runs should be constructed.

# Heavy Sentence

**4 Players**
**4 Calculators**
**Paper and Pencil**

## OBJECT OF THE GAME

For each player to use the ALPHABET CHART to generate as big a number in his calculator as is possible. The players cannot go over 15,000 per round.

## THE PLAY

The players each write a sentence on a piece of paper. The sentence must have six, seven, or eight words. The first player shows his or her sentence to the other players and they each try to get as large a value for the sentence as they can, using the ALPHABET CHART and the coordinates of each of the letters.

Each letter on the chart has twenty-six different coordinates (column number and row number). For example, look at the A in the upper left hand corner. Its coordinates are 1 (across) and 1 (down). It also has the coordinates 26 (across) and 2 (down), and it has twenty-four other combinations of coordinates. Using any operation ($+ - \times \div$) the players are to use the letters of the sentence to determine the coordinates that will give the highest total to the player. For example: if the first word of the sentence is IN, one player can use $16 \times 20 = 320$ to get I, while another player can use $14 \times 22 = 308$ to get I. Each player is trying to get the highest number (the value of the letter is ADDED TO THE VALUE OF THE NEXT LETTER but no player can go over 15,000 for one sentence). The player who gets the highest number without going over

15,000 wins that round. There are four rounds, one for each sentence. Whoever wins the most rounds wins the game.

## STRATEGY

It is probably best to build your number as fast as possible up to a certain point and then slow down. If you go over 15,000 you lose, so be careful after you pass the middle of the sentence.

## SAMPLE PLAY

*Player A's sentence:* "We are trying to get a high number."

*Player A*

For W 26 × 24 = 624, for E 26 × 6 = 156, for A 16 × 12 = 192, for R 26 × 19 = 494, for E 20 × 12 = 240 (Player A has "we are" for 1,706 points).

*Player B*

For W 25 × 25 = 625, for E 18 × 14 = 252, for A 18 × 10 = 180, for R 22 × 23 = 506, for E 18 × 14 = 252 (Player B has "we are" for 1,815 points).

*Player C*

For W 14 × 10 = 140, for E 19 × 13 = 247, for A 16 × 12 = 192, for R 24 × 21 = 504, for E 19 × 13 = 247 (Player C has "we are" for 1,301 points).

*Player D*

For W 24 × 26 = 624, for E 16 × 16 = 256, for A 17 × 11 = 187, for R 21 × 24 = 504, for E 16 × 16 = 256 (Player D has "we are" for 1,827 points).

Player D is leading after five letters.

## COMPLEX VERSION

The rounds are limited to ten minutes each.

## ALPHABET CHART

|    | 1 | 2 | 3 | 4 | 5 | 6 | 7 | 8 | 9 | 10 | 11 | 12 | 13 | 14 | 15 | 16 | 17 | 18 | 19 | 20 | 21 | 22 | 23 | 24 | 25 | 26 |
|----|---|---|---|---|---|---|---|---|---|----|----|----|----|----|----|----|----|----|----|----|----|----|----|----|----|----|
| 1  | A | B | C | D | E | F | G | H | I | J | K | L | M | N | O | P | Q | R | S | T | U | V | W | X | Y | Z |
| 2  | B | C | D | E | F | G | H | I | J | K | L | M | N | O | P | Q | R | S | T | U | V | W | X | Y | Z | A |
| 3  | C | D | E | F | G | H | I | J | K | L | M | N | O | P | Q | R | S | T | U | V | W | X | Y | Z | A | B |
| 4  | D | E | F | G | H | I | J | K | L | M | N | O | P | Q | R | S | T | U | V | W | X | Y | Z | A | B | C |
| 5  | E | F | G | H | I | J | K | L | M | N | O | P | Q | R | S | T | U | V | W | X | Y | Z | A | B | C | D |
| 6  | F | G | H | I | J | K | L | M | N | O | P | Q | R | S | T | U | V | W | X | Y | Z | A | B | C | D | E |
| 7  | G | H | I | J | K | L | M | N | O | P | Q | R | S | T | U | V | W | X | Y | Z | A | B | C | D | E | F |
| 8  | H | I | J | K | L | M | N | O | P | Q | R | S | T | U | V | W | X | Y | Z | A | B | C | D | E | F | G |
| 9  | I | J | K | L | M | N | O | P | Q | R | S | T | U | V | W | X | Y | Z | A | B | C | D | E | F | G | H |
| 10 | J | K | L | M | N | O | P | Q | R | S | T | U | V | W | X | Y | Z | A | B | C | D | E | F | G | H | I |
| 11 | K | L | M | N | O | P | Q | R | S | T | U | V | W | X | Y | Z | A | B | C | D | E | F | G | H | I | J |
| 12 | L | M | N | O | P | Q | R | S | T | U | V | W | X | Y | Z | A | B | C | D | E | F | G | H | I | J | K |
| 13 | M | N | O | P | Q | R | S | T | U | V | W | X | Y | Z | A | B | C | D | E | F | G | H | I | J | K | L |
| 14 | N | O | P | Q | R | S | T | U | V | W | X | Y | Z | A | B | C | D | E | F | G | H | I | J | K | L | M |
| 15 | O | P | Q | R | S | T | U | V | W | X | Y | Z | A | B | C | D | E | F | G | H | I | J | K | L | M | N |
| 16 | P | Q | R | S | T | U | V | W | X | Y | Z | A | B | C | D | E | F | G | H | I | J | K | L | M | N | O |
| 17 | Q | R | S | T | U | V | W | X | Y | Z | A | B | C | D | E | F | G | H | I | J | K | L | M | N | O | P |
| 18 | R | S | T | U | V | W | X | Y | Z | A | B | C | D | E | F | G | H | I | J | K | L | M | N | O | P | Q |
| 19 | S | T | U | V | W | X | Y | Z | A | B | C | D | E | F | G | H | I | J | K | L | M | N | O | P | Q | R |
| 20 | T | U | V | W | X | Y | Z | A | B | C | D | E | F | G | H | I | J | K | L | M | N | O | P | Q | R | S |
| 21 | U | V | W | X | Y | Z | A | B | C | D | E | F | G | H | I | J | K | L | M | N | O | P | Q | R | S | T |
| 22 | V | W | X | Y | Z | A | B | C | D | E | F | G | H | I | J | K | L | M | N | O | P | Q | R | S | T | U |
| 23 | W | X | Y | Z | A | B | C | D | E | F | G | H | I | J | K | L | M | N | O | P | Q | R | S | T | U | V |
| 24 | X | Y | Z | A | B | C | D | E | F | G | H | I | J | K | L | M | N | O | P | Q | R | S | T | U | V | W |
| 25 | Y | Z | A | B | C | D | E | F | G | H | I | J | K | L | M | N | O | P | Q | R | S | T | U | V | W | X |
| 26 | Z | A | B | C | D | E | F | G | H | I | J | K | L | M | N | O | P | Q | R | S | T | U | V | W | X | Y |

# The Board Game

**2 or more Players**
**1 Calculator per Player**
**1 Die**
**1 Chip per Player**
**Paper and Pencil**

## OBJECT OF THE GAME

This is a game with two possible objects:

(1) Cooperative object: players try to land on the same square in the fewest number of turns.

(2) Competitive object: player who can move into a square occupied by another player wins.

## THE PLAY

Using a regular chess or checkers board (or drawing a board on paper) mark or think of the squares as having the following numbers associated with them. Only the outside ring of squares is numbered, and the players must stay on these squares.

| 1 | 7 | 6 | 5 | 4 | 3 | 2 | 1 |
|---|---|---|---|---|---|---|---|
| 2 |   |   |   |   |   |   | 7 |
| 3 |   |   |   |   |   |   | 6 |
| 4 |   |   |   |   |   |   | 5 |
| 5 |   |   |   |   |   |   | 4 |
| 6 |   |   |   |   |   |   | 3 |
| 7 |   |   |   |   |   |   | 2 |
| 1 | 2 | 3 | 4 | 5 | 6 | 7 | 1 |

Each player has a chip to mark his or her position on the board. At the beginning of the game, the chips are placed on "1" square. The starting player rolls a single die and moves a distance equal to the number that comes up plus the number of the square on which his chip presently lies. Thus, starting on square 1 and rolling a 5, the first player would move $5 + 1 = 6$, in either direction. The second player moves in a similar fashion (starting at 1 and rolling a 2, he or she would move $1 + 2 = 3$). Succeeding moves are made in the same manner. If Player A rolls a 4 on the second turn, he or she moves either $7 + 4 = 11$ or $2 + 4 = 6$, depending on the direction of the original move.

For the cooperative version, players try to land on the same square in the fewest number of moves. In the competitive version, the player who can move onto the square occupied by the other player's chip wins.

## SAMPLE PLAY

| PLAYER | POSITION | COOPERATIVE DIE | MOVE | NOW ON SQUARE |
|---|---|---|---|---|
| A (top right) | 1 | 5 | 6 | 2 |
| B (bottom right) | 1 | 6 | 7 | 1 |
| C (bottom left) | 1 | 3 | 4 | 4 |
| D (top left) | 1 | 2 | 3 | 5 |
| A | 2 | 5 | 7 | 2 |
| B | 1 | 3 | 4 | 4 (B and C are on same square and have won. A and D continue until they are on same square) |

Notice that all the players went clockwise around the board. Either player could have gone counterclockwise at any time.

## COMPLEX VERSION

In this game, players roll a die and move according to the die number and their present position on the board. With each board position number there is associated an action which tells the number of squares a player must move. The list follows:

| SQUARE NUMBER | FUNCTION |
|---|---|
| 1 | Die number plus opponent's position |
| 2 | Twice the die number |
| 3 | Twice the die number minus 2 |
| 4 | Twice opponent's position minus die number |
| 5 | Twice die number minus opponent's position |
| 6 | Die number $\times$ opponent's number divided by 2 |
| 7 | Die number $\times$ your position divided by 2 |

Ignore all fractional parts and negative signs. For example: a player on a square number 7 rolling a 3 would move $7 \times 3 \div 2 = 10$ (not $10\frac{1}{2}$) squares. A player on square 4 rolling a 2 (opponent is on 5) will move $2 \times 5 - 2 = 8$ squares.

# Multiplication Maze

1 Player
1 Calculator

## OBJECT OF THE PUZZLE

To get from the squares numbered 1 through 8 to the square numbered 90 by multiplying the two numbers in the boxes times each other. *The route does not necessarily have to be through adjacent boxes,* but is arrived at through multiplication.

## THE PLAY

For each maze, begin at the indicated START LOCATIONS (the small maze has one starting box while the large maze has eight starting boxes). In the small maze below, entering the box labeled "1," multiply 1 by the number in the lower right corner of the box (2). Thus, $1 \times 2 = 2$. 2 is the number of the next box. Moving to box "2" and multiplying, $2 \times 1.5 = 3$. Notice that there are two boxes labeled "3." This means the path splits and you must find the correct path. There may be one, two, or more branches from any box. Note that a branch need not lie adjacent to the multiplier box. The path to the finish may tunnel under some boxes.

Only one path leads to the FINISH. The correct path consists of an unbroken but not necessarily adjacent string of multiplications from the starting point to the finish (such as $1 \times 2 = 2, 2 \times 1.5 = 3, 3 \times 2 = 6$, et cetera). See if you can find the path which leads to the finish. If you do not wish to mark the book, place a piece of tracing paper over the page. Read numbers through the paper and trace paths and branches as you multiply them.

Note that if the multiplication factor is a fraction, such as 7/9, multiply by the top number and then divide by the bottom number for the new box number. (If the box is "27" and the multiplier 7/9, the operation is $27 \times 7 \div 9 = 21$.)

**SAMPLE PLAY**

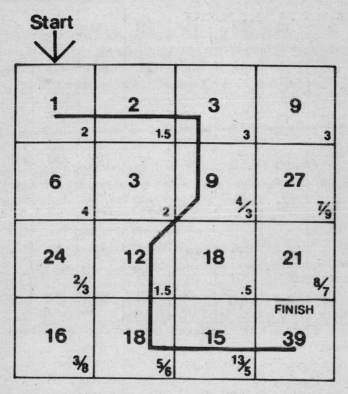

You may start at any arrow:

Start ↓ ↓ ↓  Start ↓ ↓  Start ↓ ↓

| | | | | | | | |
|---|---|---|---|---|---|---|---|
| **1** $_9$ | **2** $_5$ | **3** $_3$ | **4** $_4$ | **5** $_{1.2}$ | **6** $_{7/8}$ | **7** $_7$ | **8** $_{1/2}$ |
| **9** $_2$ | **10** $_2$ | **32** $_{.75}$ | **16** $_2$ | **60** $_{1/12}$ | **60** $_{1.2}$ | **72** $_{7/8}$ | **28** $_{.25}$ |
| **20** $_{.6}$ | **12** $_2$ | **24** $_{2/3}$ | **30** $_{1.5}$ | **45** $_{.6}$ | **54** $_{10/9}$ | **63** $_{10/9}$ | **70** $_{1.1}$ |
| **18** $_2$ | **12** $_{1.25}$ | **35** $_{6/7}$ | **45** $_{10/9}$ | **27** $_{14/9}$ | **54** $_{.5}$ | **49** $_{12/7}$ | **77** $_{6/7}$ |
| **12** $_{7/6}$ | **15** $_{1.4}$ | **50** $_{.7}$ | **48** $_{25/24}$ | **42** $_{8/7}$ | **33** $_{18/11}$ | **66** $_{5/6}$ | **84** $_{8/7}$ |
| **36** $_{19/36}$ | **14** $_{13/7}$ | **21** $_{1/.84}$ | **11** $_2$ | **22** $_{25/11}$ | **44** $_{.75}$ | **55** $_{.6}$ | **96** $_{5/6}$ |
| **26** $_{.5}$ | **42** $_{.5}$ | **19** $_1$ | **25** $_{1.6}$ | **52** $_{11/13}$ | **33** $_{8/3}$ | **55** $_{18/11}$ | **80** $_{.8}$ |
| **13** $_{11/13}$ | **56** $_{.75}$ | **40** $_{1.4}$ | **45** $_{13/15}$ | **39** $_{4/3}$ | **88** $_{9/11}$ | **64** $_{1.25}$ | FINISH **90** |

# Duo-Maze

**2 Players**
**2 Calculators**
**2 Chips**

## OBJECT OF THE GAME

This two-person maze game is similar to the Multiplication Maze. Players take turns multiplying and moving through the maze, trying to reach the goal (100).

## THE PLAY

Duo-Maze is played on the board which is shown on page 153. There are two entry boxes into the maze, one for each player. Each box has a large number in the middle (1 and 2 for the entry boxes) which is the box number. Boxes also contain numbers in the upper left and lower right corners with operations such as $+ -$ $\times \div$. The box number is acted upon by either of these numbers and operations to find the boxes which connect with it boxes that may not be adjacent. In the example below,

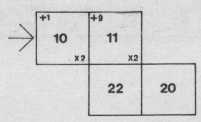

box 10 goes either to: box 10 + 1 = 11 or to box 10 × 2 = 20. Thus there is a choice of going to either box #11 or box #20. Similarly, box 11 is connected to box 11 + 9 = 20, and box 11 × 2 = 22.

Players may use either path. Note that in some instances, there may be more than one box with the same number. This means that the path branches in two additional directions.

Players begin by placing chips at their starting square. If the board in the book is too small, a chess or checkers board may be used, referring to the book for positions.

Player A begins by calculating the squares which his initial square connects with and moving into one of them. Player B then moves his token in the same manner. Players may only move to squares which are connected by the factors. *Spatial proximity plays no part* (because two boxes are side by side, this does not mean they are connected).

Players take turns advancing one square at a time. There are no blind alleys, so all paths eventually lead to the goal.

If in the course of play one player moves into a square adjacent to the square occupied by the other player, he or she will immediately get another turn (being able to advance one extra square) according to the box he or she is in. If a player moves into the same square as the other player, the player immediately gets two additional turns.

Note that if the multiplication factor is a fraction, such as 7/9, multiply by the top number and then divide by the bottom number for the new box number. (If the box is "27" and the multiplier 7/9, the operation is $27 \times 7 \div 9 = 21$.)

| | | B | | | | | |
|---|---|---|---|---|---|---|---|
| x2 · **4** · -1 | ÷2 · **6** · x⅔ | x2.5 · **2** · x3 | +1 · **5** · -1 | -2 · **80** · x.9375 | ÷.9375 · **90** · ÷1.8 | ÷1.625 · **78** · x²³⁄₂₆ | -1 · **69** · x²²⁄₂₃ |
| x2 · **3** · x3 | +33 · **9** · -1 | x7 · **6** · x1.5 | ÷15 · **75** · x1.12 | ÷1.28 · **96** · -4 | x¹⁸⁄₁₃ · **65** · ÷.8125 | x1.125 · **72** · x¹³⁄₁₂ | ÷2 · **63** · x²³⁄₂₁ |
| A · x3 · **1** · x7 | x6 · **7** · x2 | ÷2 · **42** · ÷3 | ÷4 · **84** · x²²⁄₂₁ | x²¹⁄₂₃ · **92** · ÷7 | -3 · **81** · ÷.9 | ÷7 · **49** · x⁹⁄₇ | x¹²⁄₁₁ · **66** · -10 |
| x1.5 · **8** · ÷2 | -6 · **14** · ÷2 | -1 · **21** · ÷7 | ÷3 · **21** · ÷3 | x1.125 · **88** · x.75 | x⁹⁄₁₁ · **99** · ÷.99 | +.875 · **56** · x⅞ | ÷.85 · **51** · x²²⁄₁₇ |
| +3 · **19** · +6 | +2 · **18** · +1 | +1 · **24** · x.75 | x1.25 · **20** · ÷5 | **GOAL 100** | ÷1.36 · **68** · x.75 | +4 · **64** · x.875 | x⁴⁄₃ · **48** · x⁷⁄₆ |
| ÷2 · **22** · +8 | -8 · **30** · ÷3 | x.72 · **25** · x1.2 | ÷2.75 · **55** · ÷11 | x¾ · **44** · x1.25 | +18 · **26** · ÷10 | x1.2 · **50** · x52 | x.8 · **60** · ÷1.6 |
| +4 · **12** · ÷2 | -3 · **28** · ÷2 | +17 · **11** · x3 | ÷1 · **33** · ÷2 | -9 · **35** · ÷.7 | x1 · **52** · ÷2 | -2 · **54** · -4 | -6 · **17** · +1 |
| ÷2 · **16** · x1.3125 | x1.5 · **10** · x2 | x1.8 · **15** · -4 | x⅔ · **27** · x⅓ | ÷8 · **36** · +4 | x1.3 · **40** · x.375 | -11 · **45** · x1.2 | +2 · **34** · ÷2 |

# Quartet Maze

**4 Players**
**4 Calculators**
**4 Chips**

## OBJECT OF THE GAME

The Quartet Maze is similar to the Multiplication Maze. Players take turns multiplying and moving through the maze trying to reach the goal (98).

## THE PLAY

The game is played on the 8 by 8 board shown on the next page. There are four starting boxes, one for each player. As in the Duo-Maze, the boxes each have a number, and movement from box to box is accomplished by multiplying, adding, subtracting or dividing (according to the indicated operation) the box number in the upper left or lower right corner. Players may take either path leading from a box. No two boxes have the same number. Refer to Duo-Maze for example of movement.

As in Duo-Maze, players begin by placing chips in starting boxes, taking turns calculating, and moving one square at a time. Players may only move to boxes which are connected to their present box by calculator operations, *i.e.* they move by numerical operations and not because the boxes are adjacent. There are no blind paths or boxes with factors leading to numbers not on the board.

If, in the course of play, one player moves into the square occupied by another player, the first player immediately gets one additional turn.

Note that if the multiplication factor is a fraction, such as 7/9, multiply by the top number and then divide by the bottom number for the new box number. (If the box is "27," and the multiplier 7/9, the operation is $27 \times 7 \div 9 = 21$.)

| | | | | | | | |
|---|---|---|---|---|---|---|---|
| +4 **1** x10 | x2.2 **10** x2.4 | x2 **22** +5 | +3 **11** -2 | x7 **9** +2 | -11 **25** x1.16 | ÷10.75 **86** +4 | x $43/37$ **74** x $32/37$ |
| x4.6 **5** x4 | x1.1 **20** x1.2 | ÷2 **24** x2 | x $12/11$ **44** x $15/11$ | +11 **14** -3 | +1 **63** ÷9 | x2 **29** ÷3.625 | -1 **75** ÷3 |
| x12.5 **2** x3 | x5.5 **6** x2 | -10 **33** ÷.75 | ÷9 **81** ÷2.025 | ÷1.8 **72** ÷2.88 | x.875 **64** x1.125 | ÷7.25 **58** x $31/29$ | x $7/6$ **84** ÷1.12 |
| x1.2 **25** x1.08 | x.4 **30** x1.1 | x2 **12** x4 | ÷1.08 **54** x1.5 | x6 **8** x6.5 | ÷2 **68** -4 | ÷3.6 **90** ÷1.2 | GOAL **98** |
| x5 **3** x13 | x1.8 **15** x1.2 | -5 **27** ÷.675 | x1.25 **40** x.9 | ÷1.725 **69** x $24/23$ | x $31/28$ **56** x.875 | x $28/31$ **62** x1.5 | +19 **65** -3 |
| +1.5 **39** x $11/13$ | +1 **26** x $9/13$ | -2 **18** x2 | +1.5 **60** x.9 | +1.15 **46** x1.5 | ÷7 **49** ÷.7 | -5 **93** ÷1.86 | +9.5 **76** x $21/19$ |
| x7 **4** x8 | -9 **32** ÷2 | +10 **23** ÷.575 | +.96 **48** x1.25 | +.175 **7** x3 | ÷11 **88** x.875 | x $9/7$ **70** x $11/11$ | +2 **66** x $38/33$ |
| x.75 **28** x $11/14$ | ÷1 **21** ÷1.3125 | +2 **16** x2.25 | +.9 **36** ÷.72 | -2 **50** x.92 | ÷.9625 **77** ÷1.4 | x1.2 **55** ÷6.875 | x.8125 **80** x1.125 |

# INDEX OF GAMES

# The Authors

Edwin Schlossberg received Ph.D. degrees in Science and Literature from Columbia University. He has taught at Columbia University, M.I.T., and the University of Illinois, and designed the Learning Environment for the Brooklyn Children's Museum. He is the author of *Wordswordswords, Einstein and Beckett,* and coauthor of *Projex.* Dr. Schlossberg lives in Chester, Massachusetts.

John Brockman is the author of several works of contemporary philosophy. His books, *By the Late John Brockman, 37,* and *Afterwords,* are the subject of a collection of essays, *After Brockman.* Mr. Brockman is the editor of *Real Time 1* and *Real Time 2.* He lives in New York City.

# RULES OF THE GAME

*"This book offers the best means for quickly finding a basic rule,
either disputed or forgotten, in almost any sport. What other work
can claim as much?"*

*Olympic Review*

Just in time for the Summer Olympics Corgi publish for the first
time in paperback the only comprehensive, illustrated guide to
understanding more than four hundred national and international
sporting events. To help the reader fully appreciate the Olympic
events those sports that appear in the Olympic calendar carry a
symbol beneath their title.

More than one hundred and fifty sports and four hundred sporting
activities are fully covered. Sports of a similar nature are grouped
together; each group of sports are first presented in the contents
page and later each particular sporting activity appears in its
alphabetical sequence so that answers to sport questions are fast and
easy to find.

The information is presented in concise, clearly identified sections
of copy with two thousand five hundred illustrations in full colour to
provide the reader with an easy to find reference work.

SBN 0 552 9800 2 — Price £2.95